T0201022

Pulsed EM Field Computation in Planar Circuits

The Contour Integral Method

Martin Štumpf

Brno University of Technology
Brno, Czech Republic

CRC Press
Taylor & Francis Group
Boca Raton London New York

CRC Press is an imprint of the
Taylor & Francis Group, an **informa** business

A SCIENCE PUBLISHERS BOOK

CRC Press
Taylor & Francis Group
6000 Broken Sound Parkway NW, Suite 300
Boca Raton, FL 33487-2742

First issued in paperback 2021

© 2018 by Taylor & Francis Group, LLC
CRC Press is an imprint of Taylor & Francis Group, an Informa business

No claim to original U.S. Government works

Version Date: 20180427

ISBN-13: 978-0-367-78120-0 (pbk)
ISBN-13: 978-1-138-73524-8 (hbk)

This book contains information obtained from authentic and highly regarded sources. Reasonable efforts have been made to publish reliable data and information, but the author and publisher cannot assume responsibility for the validity of all materials or the consequences of their use. The authors and publishers have attempted to trace the copyright holders of all material reproduced in this publication and apologize to copyright holders if permission to publish in this form has not been obtained. If any copyright material has not been acknowledged please write and let us know so we may rectify in any future reprint.

Except as permitted under U.S. Copyright Law, no part of this book may be reprinted, reproduced, transmitted, or utilized in any form by any electronic, mechanical, or other means, now known or hereafter invented, including photocopying, microfilming, and recording, or in any information storage or retrieval system, without written permission from the publishers.

For permission to photocopy or use material electronically from this work, please access www.copyright.com (http://www.copyright.com/) or contact the Copyright Clearance Center, Inc. (CCC), 222 Rosewood Drive, Danvers, MA 01923, 978-750-8400. CCC is a not-for-profit organization that provides licenses and registration for a variety of users. For organizations that have been granted a photocopy license by the CCC, a separate system of payment has been arranged.

Trademark Notice: Product or corporate names may be trademarks or registered trademarks, and are used only for identification and explanation without intent to infringe.

Visit the Taylor & Francis Web site at
http://www.taylorandfrancis.com

and the CRC Press Web site at
http://www.crcpress.com

*This book is dedicated to
my loving wife, Zuzana
and to my lovely daughters
Terezka and Klárka*

Preface

The present monograph is meant to provide a unified account on the state-of-the-art of the time-domain contour integral method with its applications to analyzing planar circuits. The book is in part based on results that have been published as individual research papers. In order to translate this material into a coherent story, these works have been further supplemented with new sections, appendices and demonstrational code implementations in MATLAB®. In addition, the book contains brand-new chapters describing original approaches for analyzing pulsed electromagnetic field propagation in planar structures through contour-integral problem formulations.

While the book is almost entirely devoted to the time-domain analysis of planar structures, the intended purpose of the book is more general. In particular, it is hoped that the introduced (reciprocity-based) formulations and their numerical solutions will provide a convenient point of departure for developing novel time-domain integral-equation solution methodologies, which as yet belongs to rarely explored subjects of wavefield physics. As such, the book may be of value to research engineers, but also to students who could appreciate simple demonstrational codes accompanied with comments.

The author takes this opportunity to express his thanks to H. A. Lorentz Chair Emeritus Professor Adrianus T. de Hoop, Delft University of Technology, The Netherlands, for his helpful and stimulating discussions concerning wave field modeling, to Professor Marco Leone, Otto-von-Guericke University Magdeburg, Germany, for drawing his attention to the contour-integral method and, finally, to Professors Ioan E. Lager, Delft University of Technology, The Netherlands, Zbyněk Raida, Brno University of Technology, The Czech Republic, and Guy A. E. Vandenbosch, KU Leuven, Belgium, for their constant support.

The research work included in this book was dominantly sponsored by the Czech Science Foundation under Grant 17-05445Y. The research was also partly supported by the Czech Ministry of Education, Youth and Sports under Grant LD15005 [COST CZ (LD) Program] and under Grant LO1401 [National Sustainability Program]. This financial support is gratefully acknowledged.

Contents

Acronyms

CIM	Contour-Integral Method
EM	ElectroMagnetic
EMC	ElectroMagnetic Compatibility
EMI	ElectroMagnetic Interference
FD	Frequency Domain
FIT	Finite-Integration Technique
FSV	Feature Selective Validation
IE	Integral Equation
MOT	Marching-On in Time
PCB	Printed-Circuit Board
PCBs	Printed-Circuit Boards
PEC	Perfect Electric Conductor
PMC	Perfect Magnetic Conductor
TD	Time Domain
TD-CIM	Time-Domain Contour-Integral Method
TD-C^{2}IM	Time-Domain Compensation Contour Integral Method
UWB	Ultra Wide Band

Chapter 1

Introduction

With the ever-increasing data rates on high-speed digital interconnection structures, two major concerns can be distinguished. Firstly, the need for low-cost engineering design calls for efficient modeling methodologies that enable a proper (i.e. space-time) ElectroMagnetic (EM) characterization. Secondly, the aspect of particular importance is the interference analysis that secure such systems' proper performance as well as smooth co-existence complying with the international regulations on ElectroMagnetic Interference (EMI).

Modern high-speed multilayered Printed-Circuit Boards (PCBs) typically consist of a number of conducting parallel planes and via interconnections serving as the vertical routing between the layers (see Fig. 1.1a). Given the multi-scale nature and high complexity of PCB structures, it has proved to be computationally efficient and physically intuitive to separate the EM interactions involved into their dominant contributions that are amenable to an equivalent circuit representation. This strategy is followed in the physics-based approach (see e.g. [98]) in which the via-to-via and via-to-plane interactions are accounted for through the parallel-plane impedance and capacitors, respectively (see Fig. 1.1b). Given the fact that the via-to-plane interactions can be described (semi)-analytically in closed form [39, 147], the remaining challenging task is the efficient determination of the parallel-plane cavity impedance representing the EM propagation properties of a single planar cavity circuit.

Apart from PCBs, the parallel-plate cavity structure is an indispensable building block in microwave and antenna engineering, where it is commonly used for designing compact devices (see e.g. [11, 76]). Typical

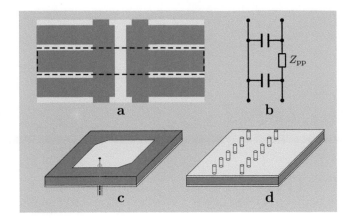

Figure 1.1: (a) Multilayered PCB and its (b) physics-based network representation; (c) Patch antenna; (d) Surface-integrated waveguide.

examples from this category are patch antennas and surface-integrated waveguides (see Figs. 1.1c and d).

A conventional numerical method capable of analyzing irregularly-shaped planar circuits in the Frequency Domain (FD) is known as the Contour-Integral Method (CIM). The CIM was pioneered by Okoshi and Miyoshi for solving various problems in microwave engineering [82]. Since its introduction at the beginning of the 1970's, the method has been successfully applied to FD modeling of planar circuits [72, 73, 80, 81, 83], microstrip antennas [43, Sec. 9.3.3] and waveguiding structures [53, 101]. The main advantages of CIM lie in its low computational demands and high versatility which enables the straightforward handling of arbitrarily-shaped planar circuits and embedding of additional circuit elements. Accordingly, CIM is still preferable in numerical modeling and optimization of complex systems where full-wave electromagnetic solvers would require exceedingly high computational resources [32, 136].

Despite the ever-increasing interest in TD modeling of high-speed interconnects [2, 97], signal integrity issues on PCBs [3, 30] or UWB antennas [35], the corresponding *time-domain* CIM has not been described in a coherent manner so far. The main purpose of this book, therefore, is to fill this gap and provide a unified description of the Time-Domain Contour-Integral Method (TD-CIM) (see [127]), with its applications to EM signal-transfer and interference analysis of planar circuits.

The standard CIM is based on the so-called cavity model [61]. Owing to its simplicity, it has been widely used at the time of onset of microstrip circuits and has rendered many valuable insights into their transmission as well as EM radiation behavior. The cavity model is based on the assumption that the EM field within the microstrip structure does not vary across the dielectric slab and is fully confined within the structure enclosed with a vanishing tangential magnetic field along its rim (magnetic wall). The vertical component of the electric field is then taken into account and the analysis is reduced to solving a two-dimensional boundary value problem that admits the closed-form eigenfunction expansion for simple shapes of the planar circuit [15, Sec. 4.14]. The assumption of the perfect magnetic wall breaks down once the height of the circuit becomes comparable to the wavelength. In such a case the effects as fringing fields or/and surface waves (if exist) may gain in significant importance and one has to resort to a full-wave technique [70]. Although the cavity model as such does not radiate any energy, radiation losses may be accounted for by introducing the equivalent dielectric loss or by imposing impedance boundary conditions along the circuit periphery [41, Sec. 2.3].

The history of CIM traces back to the 1960's and its developments are associated especially with the field of acoustics. The early origins are connected with the solution of static potential problems described in papers of Jaswon [50] and Symm [112] and later introduced by Harrington et al. [46] in electrostatics. As far as time-dependent problems are concerned, three basic approaches may be distinguished:

■ Time-stepping methods based on the finite-difference approximation of time derivatives;

■ Integral-transform methods based on the Laplace or Fourier transform;

■ Direct methods based on the time-dependent fundamental solution.

The first attempts to solve the time-dependent integral-equation formulations numerically are connected with the time-stepping approach and can be found in works of Friedman and Shaw [38] and Bennett and Weeks [7] for acoustics and electromagnetics, respectively. In their approach, a discretized integral equation (or a system of integral equations) is converted into a system of algebraic equations that is solved in a step-by-step updating manner known as the Marching-On in Time (MOT) method. The time-stepping approach uses finite-difference approximations of pertaining time derivatives. Beside others, the most popular time-integration schemes are based on the Newmark, Houbolt or Wilson θ methods [114]. Although the MOT method may suffer from serious instabilities [118, Sec. 13.3], it has become popular for tackling the transient scattering problems and its development is still ongoing (see [94], for example). The stability issues have subsequently been avoided with the aid of conjugate-gradient approaches [102, 123] and the relaxation method [117, Chapter 3], but at the expense of higher computational demands.

The second class of numerical approaches leans heavily on the application of integral transforms with respect to time. The pioneering works employing the Laplace transform are those of Cruse and Rizzo [18, 19]. One of the main disadvantages of the transform-based methods is the need for the proper choice of parameters required by inversion techniques [34, Sec. 4.2]. The corresponding integral-equation methods based on the Fourier transform can be found in [79, 115], for instance. A formulation of the direct TD integral equation method for both 2D and 3D relaxation-free scalar wave equations has been proposed by Mansur and Brebia [64, 65]. In these initial studies, a TD weighted residual form is the starting point for a numerical solution. More recent approaches are based on the convolution quadrature method [103] or on the symmetric Galerkin procedure [9]. For a detailed historical survey of the subject we refer the reader to the work of Dominguez [31].

The formulation of TD-CIM as formulated in this book is based on the reciprocity theorem of the time-convolution type (see [24, Sec. 28] and [133, Sec. 1.4.1]). In the reciprocity theorem, one of the EM-field states represents the 'actual' state while the second one is the 'testing' (or 'computational') state. The problem formulation can then be envisaged as a 'weak' form of the corresponding boundary-value problem. In contrast to the three-dimensional integral-equation formulations that require discretization of the entire surface of a conducting surface [70], the CIM accounts for its rim only, which considerably reduces the solution space and hence the computational requirements. On top of this, the corresponding space-time fundamental solution is known in closed (or semi-closed) form, which avoids the elaborate numerical evaluation of spatial inverse-Fourier integrals [71, Chapter 8]. On the other hand, the applicability of TD-CIM is limited and lies in the range where the circuit's thickness is small with respect to the spatial support of the excitation pulse. The described TD contour-integral formulations may therefore find their wide applications in the fast modeling of pulse-shaped signal transfers via parallel-plane structures, in the TD analysis of arbitrarily-shaped planar circuits and antennas and in the handling of related signal/power integrity and EMI issues.

1.1   Synopsis

In Chapter 2 a cavity model of a planar circuit is described in terms of an initial-boundary value formulation and a reciprocity-based integral relation. Both formulations are applied throughout the book to modeling the pulsed-signal transfer over planar circuits. The final section describes the incorporation of an excitation port in the reciprocity-based integral-equation formulation. This part is supplemented with Appendix A where the relevant Green-function spatial singularities are handled analytically.

Chapter 3 describes a computational method for solving the reciprocity-integral relation for the case of a loss-free, instantaneously-reacting planar circuit. In order to validate TD-CIM-based numerical results, closed-form analytical expressions are derived for a rectangular circuit. A demo MATLAB® implementation of the introduced method is briefly given in Appendix B. Finally, the results are shown to correlate very well with the ones evaluated using the (three-dimensional) Finite-Integration Technique (FIT).

The relation of the classic real-FD CIM formulation and the reciprocity relation introduced in Chapter 3 is described in Chapter 4. In this chapter, two CIM-based numerical schemes are discussed. Namely, it is shown that the classic point-matching solution can be also interpreted as a special case of the pulse-matching solution to which the 1-point Gaussian quadrature is applied. Accuracy of sample numerical results is briefly discussed with the help of a closed-form analytical formula pertaining to a rectangular planar circuit.

A topic of special importance in signal integrity on high-speed PCBs is the incorporation of dissipation and relaxation mechanisms. A semi-analytical technique for TD analysis of rectangular planar circuits with relaxation is, therefore, introduced in Chapter 5. The technique is based on a ray-like expansion and leans heavily on the numerical inversion of the Laplace transformation as detailed in Appendix I. The latter makes it possible to account for general relaxation behavior. The relation of the ray-type and the classic eigenfunction expansions is discussed. Numerical results are evaluated for two types of the dielectric relaxation function and validated, again, with the help of FIT.

In Chapter 6 the incorporation of relaxation behavior in TD-CIM is investigated. Again, the proposed approach makes use of the numerical Laplace-transform inversion introduced in Appendix I. The extension in this direction enables us to analyse arbitrarily-shaped planar circuits showing general relaxation behavior. All the obtained results are compared with the corresponding ones evaluated using the FIT.

In a number of practical applications the inclusion of lumped circuit elements is required. An example from the field of ElectroMagnetic Compatibility (EMC) is the switching-noise mitigation on high-speed PCBs using decoupling capacitors. The inclusion of basic linear lumped elements such as resistors, capacitors and inductors is hence addressed in Chapter 7. Sample numerical calculations are again confronted with FIT, thereby validating the proposed modeling technique.

In order to analyse radiation properties of microstrip antennas or EMI issues related to PCBs, the pulsed radiation characteristics of arbitrarily-shaped planar circuits are studied in Chapter 8. The comparison of TD-CIM-based and FIT-based results shows that TD-CIM may serve the purpose very well.

In Chapter 9 the computational model for efficient analysis of TD EM mutual coupling between arbitrarily-shaped planar circuits is developed with the aid of the reciprocity theorem of the time-convolution type. The interaction model makes it possible to readily evaluate the induced pulsed-voltage response of a receiving planar circuit due to the impulsive electric-current excitation applied to a transmitting planar circuit. Sample numerical results show the considerable reduction of computational demands with respect to the referential FIT.

The reciprocity theorem of the time-convolution type can also be applied to link the pulsed EM radiation characteristics to the circuit's pulsed-voltage response to a plane wave while operating in the receiving state. This is demonstrated in Chapter 10, where the relevant self-reciprocity relation concerning a general planar circuit is derived. In this chapter it is demonstrated that such a relation can be very useful for benchmarking numerical EM solvers.

The reciprocity analysis is further generalized to the N-port case in Chapter 11. In this chapter it is shown that the systematic use of the reciprocity theorem leads to the Kirchhoff-type network representation of an N-port planar circuit. The equivalent Thévenin circuit of a 2-port planar circuit is then discussed in detail. An application of the equivalent-circuit representation is finally demonstrated on sample numerical calculations of the pulsed EM radiation characteristics of a 2-port planar circuit using TD-CIM and FIT.

Chapter 12 introduces reciprocity-based closed-form expressions for the evaluation of pulsed EM-field radiated susceptibility concerning a planar circuit. Specifically, the derived relations express the pulsed voltage response of a planar structure to an external impulsive EM plane wave via a one-dimensional contour integral. In addition to the high computational efficiency of the introduced approach, the derived TD integral representations provide physical insights into the dominant (space-time) EM-coupling mechanism. Again, the formulated computational model is validated with the aid of the referential FIT.

Pulsed-EM scattering data of planar circuits may provide interesting insights into their EM behavior. Accordingly, TD EM scattering/transmitting reciprocity properties of an N-port planar circuit are studied in Chapter 13. The result is a TD compensation theorem that explicitly shows how a change of circuit's lumped loads influences the circuit's pulsed-EM scattering characteristics. The derived compensation theorem is finally validated by conducting numerical experiments using TD-CIM and FIT.

Since the EM behavior of planar circuits is frequently tuned by adding various inclusions (e.g. shorting posts, dielectric rods), an efficient TD modeling technique that would handle such inclusions is very important for practical applications. From this reason, two TD compensation theorems that address such problems are introduced in Chapter 14. In this chapter it is shown that the presence of both EM penetrable and impenetrable inclusions can be easily accounted for upon solving a space-time integral equation. The validity of the introduced concept is demonstrated on numerical calculations concerning the impact of a shorting pin in a planar circuit. These calculations are carried out with the aid of the proposed TD-CIM and the referential FIT.

In Chapter 15 the reciprocity theorem of the time-convolution type is systematically applied to formulate a new TD IE technique for analyzing planar circuits. It is demonstrated that the concept of compensation in combination with the ray-like expansion as introduced in Chapter 5 may lead to a significant reduction of the solution space and of accompanying computational costs. The drastic reduction of computational demands is demonstrated on a numerical example referring to (three-dimensional) FIT.

In Chapter 16, the compensation theorem concerning Perfect Electric Conductor (PEC) inclusions is applied in order to propose an efficient computational methodology for analyzing the impact of a set of shorting via structures on the pulsed EM-field propagation in a planar structure. The suggested methodology makes it possible to evaluate the impact of each via structure separately, which renders an illuminating tool for efficient engineering design.

1.2   Basic conventions

To localize a point in a Cartesian space $\mathbb{R}^3$, the orthogonal right-handed Cartesian reference frame is employed. The spatial reference frame is defined with respect to the origin O and the three mutually perpendicular base vectors $\{i_1, i_2, i_3\}$ of unit length each; they form in the indicated order, a right-handed system. The position vector is $x = x_1 i_1 + x_2 i_2 + x_3 i_3$. The time coordinate is denoted by t.

Except for the chapters where three-dimensional radiation characteristics and reciprocity analysis come into play (e.g. Chapters 8, 9, 10 and 12), the EM field quantities in the problem configurations are independent along the vertical direction i_3. As a consequence, it is convenient to decompose all field quantities along the horizontal plane (parallel with respect to $x_3 = 0$) and along the vertical direction. Then all symbols associated with the two-dimensional Cartesian vectors with respect to $\{i_1, i_2\}$ are typeset in bold-face Roman or bold-face Greek. The corresponding position vector, for example, would read

$$x = x_1 i_1 + x_2 i_2 \qquad (1.1)$$

Throughout the book, light-faced Roman or Greek symbols stand for scalars. Latin and Greek subscripts stand for $\{1, 2, 3\}$ and $\{1, 2\}$, respectively. The spatial differentiation with respect to x_m is denoted as ∂_m for each $m = \{1, 2, 3\}$. For example, let κ be a Cartesian vector which is differentiable with respect to spatial coordinates x_m. Then κ_μ denotes a component of κ for each $\mu = \{1, 2\}$, whose derivatives with spatial coordinates $\partial_m \kappa$ are again vector functions with components $\partial_m \kappa_\mu$. The only exceptions are ∂_t that is reserved for the partial differentiation with respect to time and ∂_ν that denotes the directional derivative along ν. The dot product and the cross product of two vectors are denoted by $\cdot$ and $\times$, respectively.

All investigated problem configurations are supposed to be linear and time invariant. To tackle such problems we preferably apply the *one-sided Laplace transformation* with respect to time accounting for the property of causality. The one-sided Laplace transformation of some bounded physical quantity is defined as

$$\hat{u}(x, s) = \int_{t=0}^{\infty} \exp(-st) u(x, t) \mathrm{d}t \qquad (1.2)$$

with $u(x, t) = 0$ for $t < 0$. Here, the transformation parameter $s \in \mathbb{C}$ (complex frequency) is chosen to have a positive real part, large enough to ensure the convergence of the Laplace integral. The corresponding time convolution of two transient space-time functions $u_1 = u_1(x, t)$ and $u_2 = u_2(x, t)$ defined on $t \in \mathbb{R}$ is given as

$$(u_1 * u_2)(x, t) = \int_{\tau \in \mathbb{R}} u_1(x, \tau) u_2(x, t - \tau) \mathrm{d}\tau$$

$$= \int_{\tau \in \mathbb{R}} u_1(x, t - \tau) u_2(x, \tau) \mathrm{d}\tau = (u_2 * u_1)(x, t) \qquad (1.3)$$

which shows the commutative properties of the time convolution. Other basic properties of the convolution are (1) linearity; (2) associative property; (3) distributive property. Applying the Laplace transformation to Eq. (1.3) yields the *convolution theorem*

$$(u_1 * u_2)(\boldsymbol{x}, t) = \mathcal{L}^{-1}[\hat{u}_1(\boldsymbol{x}, s)\hat{u}_2(\boldsymbol{x}, s)] \tag{1.4}$$

where the symbol $\mathcal{L}^{-1}(.)$ represents the inverse Laplace transformation. The convolution theorem only makes sense if there exists a strip in the complex s-plane in which the definition integrals for $\hat{u}_1(\boldsymbol{x}, s)$ and $\hat{u}_2(\boldsymbol{x}, s)$ converge simultaneously [24, Appendix B]. Finally, the time integration operator is defined as

$$\mathbf{I}_t\, u(\boldsymbol{x}, t) = \int_{\tau=-\infty}^t u(\boldsymbol{x}, \tau)\mathrm{d}\tau \tag{1.5}$$

All EM quantities are, in accordance with the international conventions, expressed in SI units (The International System of Units) [24, General Introduction].

Chapter 2

Basic Formulation

The main purpose of this chapter is to present two basic mathematical formulations concerning TD modeling of planar circuits. Namely, we describe a boundary-value formulation with the corresponding eigenfunction expansion and an integral-equation formulation based on the reciprocity theorem of the time-convolution type. As will become clear later, the former formulation is useful for validation purposes while the latter serves as the point of departure for TD-CIM.

In accordance with the definition given by Okoshi [80, Sec. 1.1.4], the planar circuit is defined here as a parallel-plane circuit whose thickness is negligible with respect to the spatial support of the excitation pulse. It is shown that under this condition, the excited EM field does not vary across the slab and the problem boils down to solving the (transverse magnetic) set of EM-field equations. The field equations, supplemented with the relevant boundary, initial and causality conditions, are then solved on a bounded surface domain. The problem may be formulated as an initial-boundary value problem and solved, when possible, with the help of conventional methods such as the separation of variable technique or the method of images. Unfortunately, application of these methods is very limited and allows analyzing circuits of elementary shapes only, such as rectangles, circles and triangles. A way to circumvent this limitation is to approach the problem via the contour-integral formulation based on the reciprocity theorem of the time-convolution type. Addressing computational implementations of the latter formulation is an objective of this book.

The present chapter is organized as follows. Firstly, the EM field equations, together with the accompanying boundary and initial conditions, are introduced. Secondly, the complex-frequency domain boundary value problem is formulated and solved in Sec. 2.1.1. Here, the solution is written out in terms of the classical eigenfunction expansion also known as the double-summation formula. In this section, we define the transmission impedance – a parameter that turns out to be proportional to a double integral of the relevant fundamental solution. Subsequently, the reciprocity-based integral-equation formulation is introduced in Sec. 2.1.2. Its relation to the boundary-value formulation is briefly outlined in the following Sec. 2.1.3. Finally, modeling strategies concerning the embedding of excitation ports are described in Sec. 2.1.4.

2.1 2D model of a planar circuit

We shall analyse the planar circuit shown in Fig. 2.1. Such a planar structure consists of a homogeneous layer that is sandwiched between two PEC planes of vanishing thickness, i.e. the upper plane Ω and the bottom plane (also called as the ground plane). The EM properties of the slab are specified by its (Boltzmann-type) dielectric relaxation function $\kappa = \kappa(t)$ and magnetic permeability $\mu = \mu_0$. Its thickness is d. The dielectric relaxation function is supposed to be causal in its EM behavior. In the case of an instantaneously reacting slab, the dielectric relaxation function is impulsive, i.e. $\kappa(t) = \epsilon\delta(t)$, and is proportional to electric permittivity ϵ. The corresponding EM wave speed is then $c = (\epsilon\mu)^{-1/2} > 0$. The structure is activated by a prescribed electric-current surface density along a section of circuit's periphery $\partial\Omega$ (i.e. boundary contour) or/and by a vertical electric-current density injected into the conducting plane Ω.

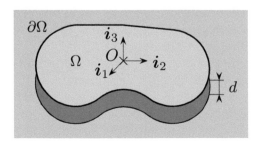

Figure 2.1: Planar circuit.

To arrive at EM field equations describing the *actual* field within the planar circuit under the thin-slab approximation, we write the field quantities in their Taylor expansions around $x_3 = 0$. After retaining only the lowest-order (x_3-independent) terms we end up with the transverse-magnetic (with respect to x_3) system of equations, i.e.

$$-\partial_1 H_2 + \partial_2 H_1 + \kappa * \partial_t E_3 = -J_3 \tag{2.1}$$

$$\partial_2 E_3 + \mu \partial_t H_1 = 0 \tag{2.2}$$

$$-\partial_1 E_3 + \mu \partial_t H_2 = 0 \tag{2.3}$$

for $x \in \Omega$ and $t > 0$. The initial values of the EM field components are assumed to be zero

$$\{E_3, \boldsymbol{H}\}(\boldsymbol{x}, 0) = \{0, \boldsymbol{0}\} \tag{2.4}$$

for all $x \in \Omega \cup \partial\Omega$. Along the circuit periphery we may prescribe the excitation electric-current surface density via the boundary-excitation condition

$$(\boldsymbol{i}_3 \times \boldsymbol{\nu}) \cdot \boldsymbol{H}(\boldsymbol{x} + \delta\boldsymbol{\nu}, t) = \boldsymbol{\nu}(\boldsymbol{x}) \cdot \partial\boldsymbol{J}(\boldsymbol{x} + \delta\boldsymbol{\nu}, t) \quad \text{as } \delta \downarrow 0 \tag{2.5}$$

for all $x \in \partial S$ and $t > 0$. Here, $\operatorname{supp}[\partial \boldsymbol{J}(\boldsymbol{x}, t)] = \partial S \subset \partial\Omega$ and $\boldsymbol{\nu}$ is the unit vector normal to $\partial\Omega$ pointing away from Ω. The classical resonator model assumes the vanishing tangential magnetic field along a source-free part of the (magnetic wall) boundary

$$(\boldsymbol{i}_3 \times \boldsymbol{\nu}) \cdot \boldsymbol{H}(\boldsymbol{x} + \delta\boldsymbol{\nu}, t) = 0 \quad \text{as } \delta \downarrow 0 \tag{2.6}$$

for all $x \in \partial\Omega/\partial S$ and $t > 0$. Equations (2.1)–(2.6) constitute an initial-boundary value problem whose solution is the main subject of the following chapters. A somewhat more general boundary condition relating the (non-zero) tangential component of the magnetic field to the vertical component of the electric field is introduced in Sec. 6.1. In the vertical electric current density J_3 introduced in Eq. (2.1) we may distinguish between the external (or active) part describing the action of the source port and the induced (or passive) part describing the current flowing through an element connected between the PEC plates of the circuit. The latter can be viewed as an equivalent contrast-source volume density producing the scattered field and is, hence, field-dependent.

The system of field equations (2.1)–(2.6) represent the starting point for developments that follow. Namely, we shall further distinguish between a boundary-value and a contour-integral formulation. While the latter will serve for modeling of arbitrarily-shaped planar circuits, the former formulation will become useful for analytical description of planar circuits having a simple shape (e.g. rectangular circuit).

2.1.1   Boundary-value formulation

Let us consider the planar structure with the perfect open boundary (cf. Eq. (2.6)) that is excited by the vertical electric-current volume density J_3. Upon applying the Laplace transformation to Eqs. (2.1)–(2.3) with (2.4) and (2.6) we may formulate the following boundary-value problem, i.e.

$$(\partial_1^2 + \partial_2^2 - \hat{\gamma}^2)\hat{E}_3 = s\mu\hat{J}_3 \quad \text{in } \Omega \tag{2.7}$$

$$\partial_\nu \hat{E}_3 = 0 \quad \text{on } \partial\Omega \tag{2.8}$$

where $\hat{\gamma} = \hat{\gamma}(s)$ is the propagation coefficient that is equal to $\hat{\gamma} = s/c$ for an instantaneously-reacting, loss-free planar circuit. Recall that ∂_ν denotes the directional derivative taken along the normal vector ν (see Sec. 1.2). The problem can be solved using Green's function which satisfies the same boundary condition along $\partial\Omega$ as the unknown electric field strength, i.e.

$$(\partial_1^2 + \partial_2^2 - \hat{\gamma}^2)\hat{G} = -\delta(\boldsymbol{x} - \boldsymbol{x}^S) \quad \text{in } \Omega \tag{2.9}$$

$$\partial_\nu \hat{G} = 0 \quad \text{on } \partial\Omega \tag{2.10}$$

which implies the linear relation between the source and the vertical electric field

$$\hat{E}_3(\boldsymbol{x}^S, s) = -s\mu \int_{\boldsymbol{x}\in\Omega^S} \hat{G}(\boldsymbol{x}|\boldsymbol{x}^S, s)\hat{J}_3(\boldsymbol{x}, s)\mathrm{d}A(\boldsymbol{x}) \tag{2.11}$$

for $\boldsymbol{x}^S \in \Omega$, $\Omega^S = \mathrm{supp}(\hat{J}_3) \subset \Omega$. The introduced Green's function can be represented using (a complete set of) eigenfunctions that satisfy

$$(\partial_1^2 + \partial_2^2 + k_{mn}^2)\psi_{mn} = 0 \quad \text{in } \Omega \tag{2.12}$$

$$\partial_\nu \psi_{mn} = 0 \quad \text{on } \partial\Omega \tag{2.13}$$

It can be shown that for (2.12)–(2.13) the eigenvalues k_{mn}^2 are always real-valued, the corresponding eigenfunctions can be normalized to satisfy the orthonormality condition

$$\int_{\boldsymbol{x}\in\Omega} \psi_{mp}\psi_{nq}\mathrm{d}A(\boldsymbol{x}) = \delta_{m,n}\delta_{p,q} \tag{2.14}$$

and form a complete set enabling us to expand the Green's function in a series

$$\hat{G}(\boldsymbol{x}|\boldsymbol{x}^S, s) = \sum_{m,n} \hat{A}_{mn}(s)\psi_{mn}(\boldsymbol{x}) \tag{2.15}$$

The expansion coefficients are found upon substituting Eq. (2.15) in Eq. (2.9) and integrating the result multiplied by ψ_{pq} over Ω. Taking into account the orthonormality condition we finally arrive at

$$\hat{G}(\boldsymbol{x}|\boldsymbol{x}^S, s) = \sum_{m,n} \frac{\psi_{mn}(\boldsymbol{x})\psi_{mn}(\boldsymbol{x}^S)}{\hat{\gamma}^2 + k_{mn}^2} \tag{2.16}$$

The Green-function method may be applied in order to describe pulsed EM transmission between source and receiver ports of a planar circuit. To this end, the electric current applied to the excitation port is written as

$$\hat{I}(s) = \int_{\boldsymbol{x} \in \Omega^S} \hat{J}_3(\boldsymbol{x}, s)\mathrm{d}A(\boldsymbol{x}) \tag{2.17}$$

where Ω^S is the domain occupied by the excitation port. Similarly, the probed voltage at the receiving port is expressed as

$$\hat{V}(s) = -\frac{d}{|\Omega^P|} \int_{\boldsymbol{x} \in \Omega^P} \hat{E}_3(\boldsymbol{x}, s)\mathrm{d}A(\boldsymbol{x}) \tag{2.18}$$

where Ω^P is the domain occupied by the receiving port and $|\Omega^P|$ denotes its surface area. Consequently, the probed voltage $\hat{V}(s)$ can be related to the electric-current density according to

$$\hat{V}(s) = \frac{s\mu d}{|\Omega^P|} \int_{\boldsymbol{x}^S \in \Omega^P} \mathrm{d}A(\boldsymbol{x}^S) \int_{\boldsymbol{x} \in \Omega^S} \hat{G}(\boldsymbol{x}|\boldsymbol{x}^S, s)\hat{J}_3(\boldsymbol{x}, s)\mathrm{d}A(\boldsymbol{x}) \tag{2.19}$$

from which the s-domain transfer impedance follows, i.e.

$$\hat{Z}(s) = \frac{\hat{V}(s)}{\hat{I}(s)} = \frac{s\mu d}{|\Omega^P| \cdot |\Omega^S|} \int_{\boldsymbol{x}^S \in \Omega^P} \mathrm{d}A(\boldsymbol{x}^S)$$
$$\int_{\boldsymbol{x} \in \Omega^S} \hat{G}(\boldsymbol{x}|\boldsymbol{x}^S, s)\mathrm{d}A(\boldsymbol{x}) \tag{2.20}$$

where we have assumed the constant distribution of the source electric-current density over the surface of the port. For a number of special cases eigenfunctions ψ_{mn} and, hence, the integration in Eq. (2.20) admit analytical representations. Finally, due to the action of the source at $x \in S$, the pulsed voltage at $x \in P$ follows from the time convolution of the excitation electric-current pulse with the TD counterpart of the transmission impedance, i.e.

$$\mathcal{V}(t) = \mathcal{Z}(t) * \mathcal{I}(t) \tag{2.21}$$

In special cases, the time convolution can be calculated analytically, which will later serve for validation of TD-CIM introduced in Chapter 3.

2.1.2 Reciprocity-based integral formulation

As the point of departure for the transient analysis of arbitrarily-shaped planar circuits, we take the reciprocity theorem of the time-convolution type (see [24, Sec. 28.2] and [133, Sec. 1.4.1]). For later convenience, we further proceed with our analysis in the complex-FD (see Sec. 1.2). Then, taking into the account the zero initial conditions (2.4), the actual wave field satisfies the (complex-FD) field equations (cf. Eqs. (2.1)–(2.3))

$$-\partial_1 \hat{H}_2 + \partial_2 \hat{H}_1 + s\hat{\epsilon}\hat{E}_3 = -\hat{J}_3 \tag{2.22}$$

$$\partial_2 \hat{E}_3 + s\mu \hat{H}_1 = 0 \tag{2.23}$$

$$-\partial_1 \hat{E}_3 + s\mu \hat{H}_2 = 0 \tag{2.24}$$

for all $\boldsymbol{x} \in \Omega$. To arrive at a weak formulation of the formulated problem, let us consider a causal *testing* (B) wave field that satisfies the following complex-FD equations

$$-\partial_1 \hat{H}_2^B + \partial_2 \hat{H}_1^B + s\hat{\epsilon}\hat{E}_3^B = -\hat{J}_3^B \tag{2.25}$$

$$\partial_2 \hat{E}_3^B + s\mu \hat{H}_1^B = 0 \tag{2.26}$$

$$-\partial_1 \hat{E}_3^B + s\mu \hat{H}_2^B = 0 \tag{2.27}$$

for all $\boldsymbol{x} \in \mathbb{R}^2$. Note that the contrast in the EM properties between both states is assumed to be zero and the condition of causality is replaced by the requirement of boundedness along the 'sphere at infinity' [133, Sec. 1.4.3]. Upon combining Eqs. (2.22)–(2.24) with Eqs. (2.25)–(2.27), we arrive at the *local interaction quantity*

$$\partial_1 \left(\hat{E}_3 \hat{H}_2^B - \hat{E}_3^B \hat{H}_2 \right) - \partial_2 \left(\hat{E}_3 \hat{H}_1^B - \hat{E}_3^B \hat{H}_1 \right) = \hat{E}_3 \hat{J}_3^B - \hat{E}_3^B \hat{J}_3 \tag{2.28}$$

for all $\boldsymbol{x} \in \Omega$. In the next step, the local interaction quantity is integrated over the surface of the conducting plate Ω (see Fig. 2.2) and with the aid of Gauss' theorem its *global form* is found

$$\chi_\Omega(\boldsymbol{x}^S) \int_{\boldsymbol{x}\in\Omega} \hat{E}_3(\boldsymbol{x}, s) \hat{J}_3^B(\boldsymbol{x}|\boldsymbol{x}^S, s) \mathrm{d}A(\boldsymbol{x})$$

$$- \int_{\boldsymbol{x}\in\partial\Omega} \hat{E}_3(\boldsymbol{x}, s)\boldsymbol{\nu}(\boldsymbol{x}) \cdot \partial\hat{\boldsymbol{J}}^B(\boldsymbol{x}|\boldsymbol{x}^S, s)\mathrm{d}l(\boldsymbol{x})$$

$$= \int_{\boldsymbol{x}\in\Omega} \hat{E}_3^B(\boldsymbol{x}|\boldsymbol{x}^S, s) \hat{J}_3(\boldsymbol{x}, s) \mathrm{d}A(\boldsymbol{x})$$

$$- \int_{\boldsymbol{x}\in\partial\Omega} \hat{E}_3^B(\boldsymbol{x}|\boldsymbol{x}^S, s)\boldsymbol{\nu}(\boldsymbol{x}) \cdot \partial\hat{\boldsymbol{J}}(\boldsymbol{x}, s)\mathrm{d}l(\boldsymbol{x}) \tag{2.29}$$

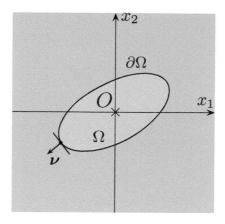

Figure 2.2: Bounded domain to which the reciprocity theorem applies.

where $\chi_\Omega(\boldsymbol{x})$ is the characteristic function $\chi_\Omega(\boldsymbol{x}) = \{1, 1/2, 0\}$ for $\boldsymbol{x} \in \{\Omega, \partial\Omega, \Omega'\}$ (Ω' denotes the complement of Ω in $\mathbb{R}^2$). Note that the second term on the right-hand side of Eq. (2.29) is zero for a planar circuit having the perfect magnetic wall along its boundary $\partial\Omega$ (see Eq. (2.6)). Following Eqs. (2.25)–(2.27), the test wave fields are linearly related to their source via

$$\hat{E}_3^B(\boldsymbol{x}|\boldsymbol{x}^S, s) =$$
$$= -s\mu \int_{\boldsymbol{x}^T \in \mathbb{R}^2} \hat{G}_\infty[r(\boldsymbol{x}|\boldsymbol{x}^T), s]\hat{J}_3^B(\boldsymbol{x}^T|\boldsymbol{x}^S, s)\mathrm{d}A(\boldsymbol{x}^T) \qquad (2.30)$$

$$\partial \hat{J}_\kappa^B(\boldsymbol{x}|\boldsymbol{x}^S, s) =$$
$$= -\int_{\boldsymbol{x}^T \in \mathbb{R}^2} \partial_\kappa \hat{G}_\infty[r(\boldsymbol{x}|\boldsymbol{x}^T), s]\hat{J}_3^B(\boldsymbol{x}^T|\boldsymbol{x}^S, s)\mathrm{d}A(\boldsymbol{x}^T) \qquad (2.31)$$

where $\hat{G}_\infty(r, s)$ is the bounded fundamental solution of the two-dimensional (modified) Helmholtz equation in $\mathbb{R}^2$ and

$$r(\boldsymbol{x}|\boldsymbol{x}^T) = |\boldsymbol{x} - \boldsymbol{x}^T| \qquad (2.32)$$

is the Eucledian distance between the points specified by (two-dimensional) position vectors $\boldsymbol{x}$ and $\boldsymbol{x}^T$. To get a boundary-contour relation, the testing surface electric current density is applied to the periphery of the circuit

$$\hat{J}_3^B(\boldsymbol{x}|\boldsymbol{x}^S, s) = \partial \hat{J}_3^B(\boldsymbol{x}|\boldsymbol{x}^S, s)\delta(\boldsymbol{x} - \boldsymbol{x}^S) \qquad (2.33)$$

where $\delta(\boldsymbol{x} - \boldsymbol{x}^S)$ being the Dirac delta distribution operative along $\boldsymbol{x}^S \in \partial\Omega$. In this way, we get the following reciprocity relation

$$\frac{1}{2} \int_{\boldsymbol{x} \in \partial\Omega} \hat{E}_3(\boldsymbol{x}, s) \partial \hat{J}_3^B(\boldsymbol{x}|\boldsymbol{x}^S, s) \mathrm{d}l(\boldsymbol{x})$$

$$- \int_{\boldsymbol{x} \in \partial\Omega} \hat{E}_3(\boldsymbol{x}, s) \boldsymbol{\nu}(\boldsymbol{x}) \cdot \partial \hat{\boldsymbol{J}}^B(\boldsymbol{x}|\boldsymbol{x}^S, s) \mathrm{d}l(\boldsymbol{x})$$

$$= \int_{\boldsymbol{x} \in \Omega} \hat{E}_3^B(\boldsymbol{x}|\boldsymbol{x}^S, s) \hat{J}_3(\boldsymbol{x}, s) \mathrm{d}A(\boldsymbol{x})$$

$$- \int_{\boldsymbol{x} \in \partial\Omega} \hat{E}_3^B(\boldsymbol{x}|\boldsymbol{x}^S, s) \boldsymbol{\nu}(\boldsymbol{x}) \cdot \partial \hat{\boldsymbol{J}}(\boldsymbol{x}, s) \mathrm{d}l(\boldsymbol{x}) \qquad (2.34)$$

with

$$\hat{E}_3^B(\boldsymbol{x}|\boldsymbol{x}^S, s) =$$

$$= -s\mu \int_{\boldsymbol{x}^T \in \partial\Omega} \hat{G}_\infty[r(\boldsymbol{x}|\boldsymbol{x}^T), s] \partial \hat{J}_3^B(\boldsymbol{x}^T|\boldsymbol{x}^S, s) \mathrm{d}l(\boldsymbol{x}^T) \qquad (2.35)$$

$$\partial \hat{J}_\kappa^B(\boldsymbol{x}|\boldsymbol{x}^S, s) =$$

$$= - \int_{\boldsymbol{x}^T \in \partial\Omega} \partial_\kappa \hat{G}_\infty[r(\boldsymbol{x}|\boldsymbol{x}^T), s] \partial \hat{J}_3^B(\boldsymbol{x}^T|\boldsymbol{x}^S, s) \mathrm{d}l(\boldsymbol{x}^T) \qquad (2.36)$$

Reciprocity relation (2.34) with Eqs. (2.35)–(2.36) serves as the basis for TD-CIM. This method yields the electric-field space-time distribution along the circuit boundary $\partial\Omega$, which is sufficient to characterize the planar circuit at hand. In particular, the superposition of the resulting field distribution along $\partial\Omega$ results in the field within circuit's domain Ω. Again, the corresponding expression directly follows from Eq. (2.29) along with the magnetic wall boundary condition (2.5), i.e.

$$\hat{E}_3(\boldsymbol{x}^S, s) = -s\mu \int_{\boldsymbol{x} \in \Omega} \hat{G}_\infty[r(\boldsymbol{x}|\boldsymbol{x}^S), s] \hat{J}_3(\boldsymbol{x}, s) \mathrm{d}A(\boldsymbol{x})$$

$$- \int_{\boldsymbol{x} \in \partial\Omega} \hat{E}_3(\boldsymbol{x}, s) \partial_\nu \hat{G}_\infty[r(\boldsymbol{x}|\boldsymbol{x}^S), s] \mathrm{d}l(\boldsymbol{x}) \qquad (2.37)$$

for $\boldsymbol{x}^S \in \Omega$, where we let $\hat{J}_3^B(\boldsymbol{x}|\boldsymbol{x}^S, s) = \delta(\boldsymbol{x} - \boldsymbol{x}^S)$. Obviously, the first term on the right-hand side can be interpreted as the primary field due to the excitation port whose action is accounted for by $\hat{J}_3$, while the second term represents the superposition of secondary contributions emanating from the circuit rim. It should be stressed that Eq. (2.37) is not an integral equation to solve, but rather a formula to be evaluated for the (known) excitation and the field distribution on $\partial\Omega$. A sample MATLAB® implementation of the field response calculation according to Eq. (2.37) is described in Sec. B.6. A relevant numerical example can be found in Sec. 6.3.2.

2.1.3   An alternative formulation

Yet another formulation that shows a relation between the boundary-value and reciprocity-based formulations may be studied. To this end, let the testing field satisfy the complex-FD field equations (2.25)–(2.27) for all $\boldsymbol{x} \in \Omega$ with the magnetic-wall boundary condition, i.e.

$$(\boldsymbol{i}_3 \times \boldsymbol{\nu}) \cdot \hat{\boldsymbol{H}}^B(\boldsymbol{x} + \delta \boldsymbol{\nu}, s) = 0 \quad \text{as } \delta \downarrow 0 \tag{2.38}$$

for all $\boldsymbol{x} \in \partial\Omega$. Further, without loss of generality, let us assume that the circuit is activated via a vertical electric-current port only, i.e. with no current injected into the circuit's rim. Taking into account that the actual field state remains the same as in Sec. 2.1, the corresponding global reciprocity relation reads (cf. Eq. (2.29))

$$\int_{\boldsymbol{x}\in\Omega} \hat{E}_3(\boldsymbol{x}, s)\hat{J}_3^B(\boldsymbol{x}|\boldsymbol{x}^S, s)\mathrm{d}A(\boldsymbol{x}) = \int_{\boldsymbol{x}\in\Omega} \hat{E}_3^B(\boldsymbol{x}|\boldsymbol{x}^S, s)\hat{J}_3(\boldsymbol{x}, s)\mathrm{d}A(\boldsymbol{x})$$
$$\tag{2.39}$$

with

$$\hat{E}_3^B(\boldsymbol{x}|\boldsymbol{x}^S, s) = -s\mu \int_{\boldsymbol{x}^T\in\Omega} \hat{G}[r(\boldsymbol{x}|\boldsymbol{x}^T), s]\hat{J}_3^B(\boldsymbol{x}^T|\boldsymbol{x}^S, s)\mathrm{d}A(\boldsymbol{x}^T) \tag{2.40}$$

$$\partial\hat{J}_\kappa^B(\boldsymbol{x}|\boldsymbol{x}^S, s) = -\int_{\boldsymbol{x}^T\in\Omega} \partial_\kappa\hat{G}[r(\boldsymbol{x}|\boldsymbol{x}^T), s]\hat{J}_3^B(\boldsymbol{x}^T|\boldsymbol{x}^S, s)\mathrm{d}A(\boldsymbol{x}^T) \tag{2.41}$$

where the Green's function satisfies the boundary-value problem defined in (2.9) and (2.10). For a spatially concentrated point source

$$\hat{J}_3^B(\boldsymbol{x}|\boldsymbol{x}^S, s) = \hat{I}^B(s)\delta(\boldsymbol{x} - \boldsymbol{x}^S) \tag{2.42}$$

with the Dirac distribution operative at $\boldsymbol{x}^S \in \Omega$ one arrives back at the field representation (2.11), provided that we invoke the condition that the resulting relation has to hold for arbitrary values of $\hat{I}^B(s)$.

2.1.4   Modeling of excitation ports

Microstrip-line feeds and vertical ports are the most common means for exciting EM fields in planar circuits. Accordingly, in this section we describe a way in which these ports can be implemented in CIM-based techniques. An example of the circuit with a microstrip excitation port on the circuit periphery (PORT 1), a vertical excitation port (PORT 2), and two observation probes (PROBE 1 and PROBE 2) are shown in Fig. 2.3.

   Let us first describe the excitation port activated by the electric-current surface density according to the excitation condition (2.5).

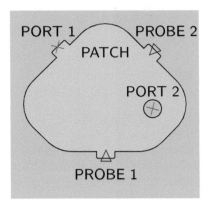

Figure 2.3: Planar circuit with its excitation ports and measurement probes.

Along (relatively small) line elements of the port's periphery, we assume the constant electric-current distribution (see Fig. 2.4). For the normal component of the excitation current injected *into* a line element $\triangle\Omega^{[P]} \subset \partial\Omega$ we write

$$\boldsymbol{\nu}(\boldsymbol{x}) \cdot \partial\hat{\boldsymbol{J}}(\boldsymbol{x}, s) = -\hat{I}(s)\Pi^{[P]}(\boldsymbol{x})/|\triangle\Omega^{[P]}| \qquad (2.43)$$

where $\hat{I}(s)$ is the electric current applied to the excitation segment, $\Pi^{[P]}(\boldsymbol{x})$ is the rectangular function defined as

$$\Pi^{[P]}(\boldsymbol{x}) = \begin{cases} 1 & \text{if } \boldsymbol{x} \in \triangle\Omega^{[P]} \\ 0 & \text{elsewhere} \end{cases} \qquad (2.44)$$

and $|\triangle\Omega^{[P]}|$ denotes the length of the excitation segment. In Eq. (2.43) one has to take care of the orientation of the injected current. In this respect it is worth noting that the right-hand side of Eq. (2.34) shows equivalence between the vertical and horizontal excitation current densities in the formulated two-dimensional model (see also [43, Sec. 9.3.1]). This implies that in the two-dimensional model it does not make sense to distinguish between the action of the vertical electric-current density J_3 and the horizontal electric-current density $-\boldsymbol{\nu} \cdot \partial\boldsymbol{J}$ injected into the circuit periphery $\partial\Omega$. Upon combining the second term on the right-hand side of Eq. (2.34) with Eqs. (2.35) and (2.43) we finally arrive at

$$\left[-s\mu\hat{I}(s)/|\triangle\Omega^{[P]}|\right] \int_{\boldsymbol{x}^T\in\partial\Omega} \partial\hat{J}_3^B(\boldsymbol{x}^T|\boldsymbol{x}^S, s)$$

$$\int_{\boldsymbol{x}\in\triangle\Omega^{[P]}} \hat{G}_\infty[r(\boldsymbol{x}|\boldsymbol{x}^T), s]\mathrm{d}l(\boldsymbol{x})\mathrm{d}l(\boldsymbol{x}^T) \qquad (2.45)$$

The integrals in Eq. (2.45) do not present any difficulties except for the overlapping discretization segments where the Green's function shows the logarithmic singularity at $\boldsymbol{x} = \boldsymbol{x}^T$. In such a case, one can use the generic integral given in Appendix A to evaluate the inner integral. Since the testing source density $\hat{J}_3^B$ is assumed to be a piecewise linear function composed of $T^{[m]}(\boldsymbol{x})$ (see Fig. 2.4), the outer integral over $\boldsymbol{x}^T$ follows easily.

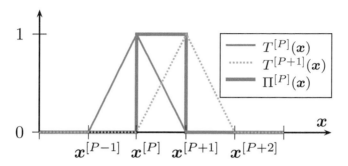

Figure 2.4: Modeling of the injected excitation current over a line element.

Alternatively, one may avoid the singularity by assuming a vertical port of the circular cross-section whose boundary contour does not belong to $\partial\Omega$. The starting point now is the first term on the right-hand side of Eq. (2.34) with Eq. (2.35), i.e.

$$(-s\mu/2\pi) \int_{\boldsymbol{x}^T \in \partial\Omega} \partial \hat{J}_3^B(\boldsymbol{x}^T | \boldsymbol{x}^S, s)$$

$$\int_{\boldsymbol{x} \in \Omega^Q} \hat{J}_3(\boldsymbol{x}, s) \mathrm{K}_0[\hat{\gamma}(s) r(\boldsymbol{x} | \boldsymbol{x}^T)] \mathrm{d}A(\boldsymbol{x}) \mathrm{d}l(\boldsymbol{x}^T) \qquad (2.46)$$

where $\mathrm{K}_0(x)$ is the modified Bessel function of the second kind and the zeroth order that represents the (bounded) fundamental solution of the modified Helmholtz equation in $\mathbb{R}^2$ [69, Sec. 11.2] (see also Sec. 2.1.1). In Eq. (2.46), $\Omega^Q = \mathrm{supp}[\hat{J}_3(\boldsymbol{x}, s)] \subset \Omega$ is the support of the excitation vertical electric-current volume density. Now, if we assume that Ω^Q is a circular domain whose radius ρ is sufficiently small with respect to the spatial support of the excitation electric-current pulse, we may approximately take

$$\hat{J}_3(\boldsymbol{x}, s) \simeq \hat{I}(s)/\pi\rho^2 \quad \text{for } \boldsymbol{x} \in \Omega^Q \qquad (2.47)$$

Consequently, with the aid of the addition theorems for Bessel functions (see [111, Sec. 6.11] and [1, (9.6.3), (9.6.4)]), the inner integral can be evaluated analytically, i.e.

$$\frac{1}{\pi\rho^2}\frac{1}{2\pi}\int_{\boldsymbol{x}\in\Omega^Q}\mathrm{K}_0[\hat{\gamma}(s)r(\boldsymbol{x}|\boldsymbol{x}^T)]\mathrm{d}A(\boldsymbol{x})$$
$$=\frac{1}{\pi}\frac{1}{\hat{\gamma}(s)\rho}\mathrm{I}_1[\hat{\gamma}(s)\rho]\mathrm{K}_0[\hat{\gamma}(s)r(\boldsymbol{x}^C|\boldsymbol{x}^T)] \qquad (2.48)$$

where $\boldsymbol{x}^T \in \partial\Omega$ and $\boldsymbol{x}^C \in \Omega$ describes the center of the circular excitation-port domain. Upon collecting the results and using the limit $\lim_{x\downarrow0}\mathrm{I}_1(x)/x = 1/2$, the interaction integral (2.46) can be simplified to

$$\left[-s\mu\hat{I}(s)/2\pi\right]\int_{\boldsymbol{x}^T\in\partial\Omega}\partial\hat{J}_3^B(\boldsymbol{x}^T|\boldsymbol{x}^S,s)\mathrm{K}_0[\hat{\gamma}(s)r(\boldsymbol{x}^C|\boldsymbol{x}^T)]\mathrm{d}l(\boldsymbol{x}^T) \quad (2.49)$$

Since the circular port is placed in domain Ω occupied by the conducting plates, it is clear that the logarithmic singularity is avoided in Eq. (2.49). The second way, on the other hand, is limited to the excitation ports of the circular cross-section. Illustrative MATLAB® implementations of the described TD-CIM excitation ports can be found in Sec. B.3.

2.2   Conclusions

It has been demonstrated that pulsed EM characteristics of a planar circuit can be evaluated with the aid of the eigenfunction-expansion method and the reciprocity-based contour-integral formulation. Since the former approach yields closed-form analytical solutions for generic circuit shapes, this method is suitable for validating computational techniques such as TD-CIM formulated in this book. Owing to the fact that any numerical modeling of EM field excitation mechanisms has the decisive impact on the proper evaluation of a planar circuit, a great deal of attention has been paid to the embedding of its excitation ports. The introduced formulations provide a solid basis for the subsequent chapter where TD-CIM is described in detail.

Chapter 3

Instantaneously-reacting Planar Circuits

[1] The present chapter provides a numerical procedure for solving the contour-integral reciprocity relation in TD for the case of a loss-free, instantaneously-reacting (or dispersion free) planar circuit [127]. Without loss of generality, we shall assume that the planar circuit is excited by the electric-current surface density injected into its periphery. Owing to the equivalence noted in Sec. 2.1.4, the vertical excitation port may be modeled along the same lines. Since the loss-free section of the circuit's rim constitutes the perfect magnetic wall with the vanishing tangential component of the magnetic field strength, such a structure cannot in principle radiate and may be viewed as a closed resonator.

It turns out that the proposed approach leads to a system of algebraic equations that are solvable in an updating step-by-step manner. Within the tested input parameters, it was observed that the resulting numerical scheme is stable, provided that the relevant matrix elements are evaluated accurately enough to prevent error-accumulation instabilities (see e.g. [118]). The proposed technique enables us to investigate the pulse-shaped signal transfer between source and receiver ports placed along the periphery of an arbitrarily-shaped planar circuit. In contrast to the standard real-FD CIM that requires the matrix inverse at a large number of frequency points in combination with the inverse fast Fourier transform, the introduced marching on-in-time scheme fur-

---

[1] © [2014] IEEE. This chapter is in part adapted, with permission, from [127].

nishes the stable TD field response in the given time window via a single matrix inversion of a time-independent matrix.

The following sections are organized as follows. The chapter begins with the numerical solution of the reciprocity-based contour-integral relation. In order to validate the numerical results, an analytical closed-form solution is constructed for a rectangular circuit in Sec. 3.2. The closed-form solution is based on the eigenfunction expansion, as discussed in Sec. 2.1.1. Finally, sample numerical calculations are presented in Sec. 3.3. Here, the introduced computational procedure is validated using the analytical eigenfunction-based expressions and with the aid of FIT.

3.1 Numerical solution of the reciprocity formulation

With the aid of the (bounded) fundamental solution of the modified Helmholtz equation in the entire $\mathbb{R}^2$ space (see [69, Sec. 11.2], for instance), Eqs. (2.34)–(2.36) lead to

$$
\int_{\boldsymbol{x}\in\partial\Omega} \hat{E}_3(\boldsymbol{x},s)\partial\hat{J}_3^B(\boldsymbol{x}|\boldsymbol{x}^S,s)\mathrm{d}l(\boldsymbol{x})
$$

$$
= (s/c\pi)\int_{\boldsymbol{x}\in\partial\Omega} \hat{E}_3(\boldsymbol{x},s)\int_{\boldsymbol{x}^T\in\partial\Omega} \mathrm{K}_1\left[sr(\boldsymbol{x}|\boldsymbol{x}^T)/c\right]
$$
$$
\partial\hat{J}_3^B(\boldsymbol{x}^T|\boldsymbol{x}^S,s)\cos[\theta(\boldsymbol{x}|\boldsymbol{x}^T)]\mathrm{d}l(\boldsymbol{x}^T)\mathrm{d}l(\boldsymbol{x})
$$

$$
+(s\mu/\pi)\int_{\boldsymbol{x}\in\partial\mathcal{S}} \boldsymbol{\nu}(\boldsymbol{x})\cdot\partial\hat{\boldsymbol{J}}(\boldsymbol{x},s)\int_{\boldsymbol{x}^T\in\partial\Omega} \mathrm{K}_0\left[sr(\boldsymbol{x}|\boldsymbol{x}^T)/c\right]
$$
$$
\partial\hat{J}_3^B(\boldsymbol{x}^T|\boldsymbol{x}^S,s)\mathrm{d}l(\boldsymbol{x}^T)\mathrm{d}l(\boldsymbol{x}) \tag{3.1}
$$

for $\boldsymbol{x}^S \in \partial\Omega$, with $\partial\mathcal{S} \subset \partial\Omega$ and $\cos(\theta) = \partial_\nu r$. Although the reciprocity-based relation (3.1) will be numerically solved in TD, it is convenient to carry out the next few steps in complex-FD. As for the problem discretization, the time coordinate $\{t \in \mathbb{R}; t > 0\}$ is discretized in NT instants with the constant time step $\triangle t$

$$
\mathcal{T} = \{t_k \in \mathbb{R}; t_k = k\triangle t, \triangle t > 0, k = 1, ..., NT\} \tag{3.2}
$$

and the circuit periphery $\partial\Omega$ is discretized into N disjoint line segments

$$
\partial\Omega \simeq \cup_{m=1}^N \triangle\Omega^{[m]} \tag{3.3}
$$

and $|\triangle\Omega^{[m]}| = |\boldsymbol{x}^{[m+1]} - \boldsymbol{x}^{[m]}|$ is the length of the m-th segment and $\boldsymbol{x}^{[m]}$ is the position vector of the m-th discretization node. The approximation of the circuit periphery by straight-line segments is shown in

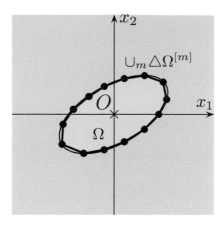

Figure 3.1: Boundary contour and its approximation by straight-line segments. © [2014] IEEE. Reprinted, with permission, from [127].

Fig. 3.1. In the following step, the electric field strength is expanded in a piecewise linear manner both in space and time, that is

$$\hat{E}_3(\boldsymbol{x}, s) = \sum_{m=1}^{N} \sum_{k=1}^{NT} e_{[k]}^{[m]} T^{[m]}(\boldsymbol{x}) \hat{T}_{[k]}(s) \tag{3.4}$$

where $e_{[k]}^{[m]}$ is the unknown coefficient, $T^{[m]}(\boldsymbol{x})$ is the triangular function associated with the m-th node along $\partial\Omega$

$$T^{[m]}(\boldsymbol{x}) = \begin{cases} |\boldsymbol{x} - \boldsymbol{x}^{[m-1]}|/|\triangle\Omega^{[m-1]}| & \text{if } \boldsymbol{x}^{[m-1]} \leq \boldsymbol{x} \leq \boldsymbol{x}^{[m]} \\ |\boldsymbol{x}^{[m+1]} - \boldsymbol{x}|/|\triangle\Omega^{[m]}| & \text{if } \boldsymbol{x}^{[m]} \leq \boldsymbol{x} \leq \boldsymbol{x}^{[m+1]} \end{cases} \tag{3.5}$$

for all $m = \{1, \cdots, N\}$ and $\hat{T}_{[k]}(s)$ is the complex-FD counterpart of the temporal triangular function (see Appendix G.1)

$$T_{[k]}(t) = \begin{cases} 1 - k + t/\triangle t & \text{if } k - 1 \leq t/\triangle t \leq k \\ 1 + k - t/\triangle t & \text{if } k \leq t/\triangle t \leq k + 1 \end{cases} \tag{3.6}$$

for all $k = \{1, \cdots, NT\}$. The testing surface current density is chosen to be piecewise linear in space and its complex-FD counterpart at the S-th node is

$$\partial \hat{J}_3^B(\boldsymbol{x}|\boldsymbol{x}^S, s) = \hat{j}_3^{[S]}(s) T^{[S]}(\boldsymbol{x}) \tag{3.7}$$

for all $S = \{1, \cdots, N\}$ and where we take $\hat{j}_3^{[S]}(s) = 1$, which corresponds to the temporal 'point matching'. The injected electric-current surface density is assumed to be constant over segments modeling the excitation port. By substituting Eqs. (3.4) and (3.7) into a discretized form of Eq. (3.1) we end up with the system of algebraic equations

$$\left(I - Q_{[0]}\right) \cdot E_{[p]} = \sum_{k=1}^{p-1} Q_{[p-k]} \cdot E_{[k]} + F_{[p]} \tag{3.8}$$

that can be solved in a step-by-step manner for all $p = \{1, \cdots, NT\}$. In Eq. (3.8), $E_{[p]}$ is a 2D-array of $[N \times NT]$ unknown coefficients at $t_p = p\triangle t$, I is a three-diagonal $[N \times N]$ 2D-array with elements

$$(I)_{S,m} = \left(\triangle\Omega^{[S-1]}/6\right)\delta_{S-1,m}$$
$$+ \left(\triangle\Omega^{[S-1]}/3 + \triangle\Omega^{[S]}/3\right)\delta_{S,m} + \left(\triangle\Omega^{[S]}/6\right)\delta_{S+1,m} \tag{3.9}$$

and $Q_{[p-k]}$ is a time-dependent $[N \times N \times NT]$ 3D-array whose elements are given as

$$(Q_{[p-k]})_{S,m} = \frac{1}{\pi c\triangle t}\int_{x^T\in\partial\Omega} T^{[S]}(x^T)\int_{x\in\partial\Omega} T^{[m]}(x)$$
$$\Psi[r(x|x^T),(p-k)\triangle t]\cos[\theta(x|x^T)]\mathrm{d}l(x)\mathrm{d}l(x^T) \tag{3.10}$$

for all $S = \{1, \cdots, N\}$, $m = \{1, \cdots, N\}$ and $t \in \mathcal{T}$. The excitation of the circuit is described via an $[N \times NT]$ 2D-array $F_{[p]}$ whose elements read

$$(F_{[p]})_S = \frac{1}{\pi|\partial S|}\int_{x^T\in\partial\Omega} T^{[S]}(x^T)$$
$$\int_{x\in\partial S} \Phi[r(x|x^T),p\triangle t]\mathrm{d}l(x)\mathrm{d}l(x^T) \tag{3.11}$$

for all $S = \{1, \cdots, N\}$ and all $t \in \mathcal{T}$, where $|\partial S|$ is the arc length of $\partial S \subset \partial\Omega$. The time-dependent functions in Eqs. (3.10) and (3.11) then follow

$$\Psi(r,t) = \psi(r,t+\triangle t) - 2\psi(r,t) + \psi(r,t-\triangle t) \tag{3.12}$$

$$\psi(r,t) = (c^2t^2/r^2 - 1)^{1/2}\mathrm{H}(t - r/c) \tag{3.13}$$

$$\Phi(r,t) = -\mu\partial_t\mathcal{I}(t) * (t^2 - r^2/c^2)^{-1/2}\mathrm{H}(t - r/c) \tag{3.14}$$

where $\mathcal{I}(t)$ is the source signature of the electric current injected into a section of the circuit periphery $\partial\Omega$ and $\mathcal{I}(t) = 0$ for $t < 0$. Evidently the source signature must be smooth enough to get an integrable singularity in Eq. (3.14) as $r \downarrow 0$. For piecewise constant/linear excitation signatures such as the rectangular or trapezoidal pulses, the time-integrated equivalent of the reciprocity relation (3.1) seems to be the shortest way to circumvent the limitation. The spatial singularity as $\boldsymbol{x}$ approaches $\boldsymbol{x}^T$ must be carefully handled via the limiting analytical procedure [140]. To this end, it is convenient to start over with Eq. (3.1) and use a small-argument expansion of the modified Bessel function [1, Eq. (9.6.13)], that is

$$\mathrm{K}_0(x) = \ln(2/x) - \gamma + \mathcal{O}(x^2) \text{ as } x \downarrow 0 \qquad (3.15)$$

where γ is Euler's constant. The corresponding integrals of the logarithmic function and its normal derivative are evaluated in Appendix A. As is shown, the latter integral gives a zero contribution for the self-coupling terms in Eq. (3.10). Since the 2D-arrays on the left-hand side of Eq. (3.8) do not depend on time, the matrix inversion is needed only once. Moreover, if $c\triangle t < \min_{S,m}[r(\boldsymbol{x}|\boldsymbol{x}^T)]$, then all elements of $\boldsymbol{Q}_{[0]}$ are zero and only the three-diagonal matrix inversion is required. For this case, efficient algorithms do exist (e.g. [47, Sec. 1.5]). More details on implementation aspects along with the corresponding demo MATLAB® codes can be found in Appendix B.

Finally note that under certain circumstances it may happen that in the effort to keep the left-hand side of Eq. (3.8) as simple as possible, a smoother temporal expansion is required. For such a case, several suitable expansion functions are given in Appendix G. The use of the the quadratic expansion functions $B_{[k]}(t)$ (see Sec. G.2), for instance, yields

$$\begin{aligned} \Psi(r,t) = \psi(r,t+\triangle t) &- 2\psi(r,t+\triangle t/2) \\ &+ 2\psi(r,t-\triangle t/2) - \psi(r,t-\triangle t) \end{aligned} \qquad (3.16)$$

$$\begin{aligned} \psi(r,t) = (2t/\triangle t)\left(c^2 t^2/r^2 - 1\right)^{1/2} \mathrm{H}(t-r/c) \\ - (2r/c\triangle t)\ln\left[ct/r + \left(c^2 t^2/r^2 - 1\right)^{1/2}\right] \mathrm{H}(t-r/c) \end{aligned} \qquad (3.17)$$

which simply replace Eqs. (3.12) and (3.13), respectively.

3.2 Analytical solutions based on the eigenfunction expansion

The main concern of this section is to provide analytical closed-form solutions that will serve for validating TD-CIM. To this end we analyse a planar circuit of the rectangular shape for which the eigenfunction expansion (2.16) is attainable in closed form.

At first let us assume a rectangular circuit of dimensions $L \times W$ that is excited by a microstrip port placed along the circuit periphery

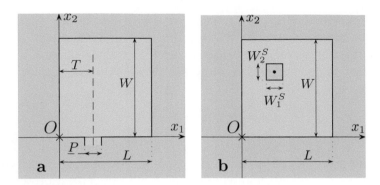

Figure 3.2: Rectangular planar circuits excited via a (a) microstrip-line source; (b) vertical-port source. © [2014] IEEE. Adapted, with permission, from [127].

$\partial\Omega^S = \{\boldsymbol{x} \in \mathbb{R}^2; T - P/2 \le x_1 \le T + P/2, x_2 = 0\}$ (see Fig. 3.2a). Then for the receiving probe of the rectangular shape $\Omega^P = \{\boldsymbol{x} \in \mathbb{R}^2; x_1^P - W_1^P/2 \le x_1 \le x_1^P + W_1^P/2, x_2^P - W_2^P/2 \le x_2 \le x_2^P + W_2^P/2\}$, Eq. (2.20) with the integrations taken over $\partial\Omega^S$ and Ω^P leads to the following complex-FD transfer impedance

$$\hat{Z}(s) = \frac{s\mu d}{LW} \sum_{m=0}^{\infty} \sum_{n=0}^{\infty} \frac{e_m^2 e_n^2}{k_m^2 + k_n^2 + s^2/c^2} F_{mn}^S F_{mn}^P \qquad (3.18)$$

where

$$F_{mn}^S = \cos(k_m T)\mathrm{sinc}(k_m P/2) \qquad (3.19)$$

$$F_{mn}^P = \cos(k_m x_1^P) \cos(k_n x_2^P)\mathrm{sinc}(k_m W_1^P/2)\mathrm{sinc}(k_n W_2^P/2) \qquad (3.20)$$

with $e_m = 1$ for $m = 0$ and $e_m = \sqrt{2}$ for $m \ne 0$ and $k_m = m\pi/L$ and $k_n = n\pi/W$. Here, the results of Sec. 2.1.1 are used. Note that a probe of vanishing dimensions can be handled via a limiting process. For a point probe, for instance, we get

$$F_{mn}^P = \cos(k_m x_1^P) \cos(k_n x_2^P) \qquad (3.21)$$

as $W_1^P \downarrow 0$ and $W_2^P \downarrow 0$. The TD counterpart of $\hat{Z}(s)$ can be found by applying the inverse Laplace transform to the terms in the sum of Eq. (3.18). With the help of [1, Eqs. (29.3.1),(29.3.16)] we find

$$\mathcal{Z}(t) = \frac{d}{LW}\frac{1}{\epsilon} \sum_{m=0}^{\infty} \sum_{n=0}^{\infty} e_m^2 e_n^2 \cos[ct(k_m^2 + k_n^2)^{1/2}]\mathrm{H}(t)F_{mn}^S F_{mn}^P \qquad (3.22)$$

As the second example, we take the rectangular circuit excited by a vertical excitation port placed on the patch Ω (see Fig. 3.2b). In such a case, the excitation port occupies a rectangular domain $\Omega^S = \{\boldsymbol{x} \in \mathbb{R}^2; x_1^S - W_1^S/2 \leq x_1 \leq x_1^S + W_1^S/2, x_2^S - W_2^S/2 \leq x_2 \leq x_2^S + W_2^S/2\}$. Again, the TD impedance follows from Eq. (3.22) with (3.20) in which F_{mn}^S is modified accordingly

$$F_{mn}^S = \cos(k_m x_1^S)\cos(k_n x_2^S)\mathrm{sinc}(k_m W_1^S/2)\mathrm{sinc}(k_n W_2^S/2) \qquad (3.23)$$

Finally, the transient voltage $\mathcal{V}(t)$ at the position of the probe, due to the impulsive source $\mathcal{I}(t)$, can be found with the help of the time convolution in Eq. (2.21).

3.3   Validation of numerical results

In this section we analyse the loss-free rectangular planar circuits whose configurations are shown in Fig. 3.2. In the both cases we consider the planar circuit of dimensions $L = 0.1\,[\mathrm{m}]$ and $W = 0.2\,[\mathrm{m}]$ with the dielectric filling of thickness $d = 1.50\,[\mathrm{mm}]$ showing the electric permittivity $\epsilon = 2.0\,\epsilon_0$ and magnetic permeability $\mu = \mu_0$. The corresponding EM wave speed in the dielectric layer is $c = (\epsilon\mu)^{-1/2}$.

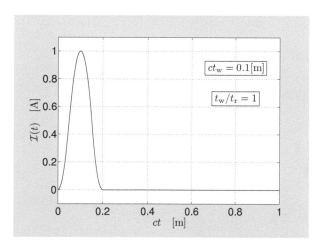

Figure 3.3: The bell-shaped excitation signature. © [2014] IEEE. Reprinted, with permission, from [127].

The circuits are excited using the bell-shaped pulse being defined in Appendix F with the amplitude $A = 1.0\,[\mathrm{A}]$ and with the pulse time width $ct_\mathrm{w} = 0.10\,[\mathrm{m}]$ (see Fig. 3.3). Note that the thin-layer assumption $ct_\mathrm{w} \gg d$ is then safely satisfied. The transient voltage responses are observed within the finite time window $\{0 \leq ct \leq 3.0\}$ [m]. For the numerical solution, the circuit periphery is divided into the line segments of length $|\triangle \Omega^{[n]}| = 0.02$ [m], which corresponds to a fifth of the excitation pulse spatial support ct_{w}. The spatial integrals in Eqs. (3.10) and (3.11) are evaluated using the 12-point Gauss-Legendre quadrature [1, (25.4.30)]. As the reference solution, we use Eq. (3.22), where the number of terms is truncated as $m = \{0, \ldots, M\}$ and $n = \{0, \ldots, N\}$, with $M = N = 1000$.

3.3.1 Rectangular circuit fed by a microstrip port

In the first example we analyse the pulse propagation over the circuit excited by the microstrip port with the parameters $T = 0.03$ [m] and $P = 0.02$ [m]. The pulsed voltage is observed along an edge of the circuit at a discretization point placed at $\{x_1^P, x_2^P\} = \{0.1, 0.16\}$ [m] with $W_1^P \downarrow 0$ and $W_2^P \downarrow 0$ (see Fig. 3.4a).

The pulse shape found via TD-CIM is compared with the analytical solution evaluated using the truncated modification of Eq. (3.22) with Eqs. (3.19), (3.21) and (2.21). The corresponding results are shown in Fig. 3.5a. Despite the coarse spatial discretization, the resulting pulse shapes correlate very well.

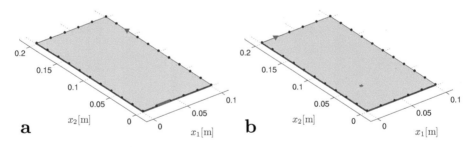

a b

Figure 3.4: Computational models of the analysed rectangular circuits with probes (the solid triangles on the circuit peripheries); (a) a microstrip port feeding (the bold line); (b) a vertical port (the dots on the patch).

3.3.2 Rectangular circuit fed by a vertical port

As the second example we observe the pulse propagation over the circuit excited by a vertical port having the rectangular cross section placed at $\{x_1^S, x_2^S\} = \{0.03, 0.05\}$ [m] with $W_1^S = W_2^S = 2.0$ [mm]. The resulting voltage response is now observed at $\{x_1^P, x_2^P\} = \{0.02, 0.2\}$ [m] (see Fig. 3.4b). In the analytical expression (3.22) we use Eqs. (3.21) and

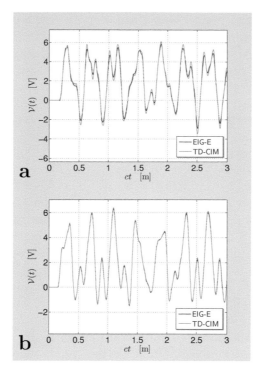

Figure 3.5: The pulsed voltage evaluated using TD-CIM and the eigenfunction expansion method (EIG-E) as observed at the (a) probe shown in Fig. 3.4a; (b) probe shown in Fig. 3.4b. © [2014] IEEE. Reprinted, with permission, from [127].

(3.23). As can be seen from Fig. 3.5b, the results are almost identical. Even better correspondence can be achieved with a finer discretization or/and with a higher number of points in the Gauss-Legendre quadrature.

3.4 Comparison with an alternative numerical technique

Another way to validate the introduced TD-CIM is to compare its results with the corresponding outputs from alternative numerical techniques. To this end, we may use FIT as implemented in CST Microwave Studio®, for example. As to the circuits excitation, we use the model of a vertical port. For the TD-CIM simulation, the port has a hexagonal cross-section of circumradius 1.0 [mm], while in the FIT-based simulation the circular port of radius 1.0 [mm] is used. The port is activated by the bell-shaped source signature plotted in Fig. 3.3. The distance between the PEC planes is again $d = 1.50$ [mm] and the electric permittivity of the loss-free dielectric filling is $\epsilon = 2.50\,\epsilon_0$. The transient voltage response is observed at a specified position on the circuit periphery within the finite time window of observation $\{0 \leq ct \leq 3.0\}$ [m]. The FIT models are discretized with a hexahedral mesh. The upper and bottom planes are assumed to be perfectly electrically conducting. The sidewalls of the surrounding box are defined as the magnetic walls with the vanishing tangential magnetic-field component.

Our first structure is of the rectangular shape and its model, as discretized for TD-CIM, is shown in Fig. 3.6a. The excitation port has its center at $\{x_1^S, x_2^S\} = \{0.025, 0.075\}$ [m]. The discretization of circuit's rim is uniform with the line segment of length $|\triangle\Omega^{[n]}| = 0.025$ [m], which corresponds to a quarter of the excitation pulse spatial support $ct_w = 0.10$ [m]. The total number of the discretization segments along the circuit periphery is 20. Our (finely-meshed) reference FIT-based model consists of about 65 thousand mesh cells. The voltage pulses observed at $\{x_1^P, x_2^P\} = \{0.1, 0.075\}$ [m] are shown in Fig. 3.7a. As the excited field quantities causally evolve over time, we can conclude that the correspondence of the early-time responses implies that our simplified excitation model corresponds well with the reference one used in the FIT-based analysis. On the other hand, the differences become evident at the late-time part of the response. This observation implies slightly different behavior of the sidewalls where the reflections take place. This can be expected due to the different models and numerical strategies used to tackle the problem.

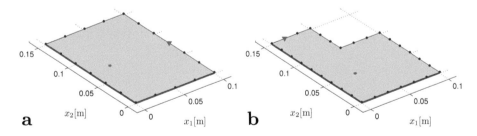

Figure 3.6: Computational models of the analysed circuits with probes (the solid triangles on the circuit peripheries) and with vertical ports (the dots on the patch); (a) the rectangular circuit; (b) the irregularly-shaped circuit.

The effect of the circuit boundary is even more pronounced in the second example shown in Fig. 3.6b. Its excitation port has its center at $\{x_1^S, x_2^S\} = \{0.03, 0.05\}$ [m]. The discretization is again uniform

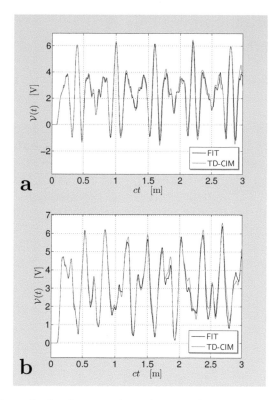

Figure 3.7: The pulsed voltage evaluated using the proposed TD-CIM and the referential FIT as observed at the (a) probe shown in Fig. 3.6a; (b) probe shown in Fig. 3.6b.

with $|\triangle\Omega^{[n]}| = 0.0167$ [m], which corresponds to one sixth of the excitation pulse spatial support $ct_w = 0.10$ [m]. The total number of the discretization segments along the circuit periphery is then 30, while the FIT model consists of 78 thousand mesh cells. The voltage pulse shapes observed at $\{x_1^P, x_2^P\} = \{0.0167, 0.15\}$ [m] are shown in Fig. 3.7b. The deviations at the late-time part of the responses can be attributed to different boundary conditions along the circuit's rim. While the CIM-based model assumes the perfect magnetic wall along the entire periphery, the magnetic wall in the FIT-based model is placed along the sidewalls of the surrounding box $x_1 = \{0, 0.10\}$ [m], $x_2 = \{0, 0.15\}$ [m] only and not along the cuts $\{0.05 \leq x_1 \leq 0.10, x_2 = 0.1\}$ [m] and $\{x_1 = 0.05, 0.10 \leq x_2 \leq 0.15\}$ [m].

3.5 Conclusions

Starting from the reciprocity-based contour-integral formulation, the Time-Domain Contour-Integral Method (TD-CIM) has been formulated. It has been demonstrated that this approach results in a stable step-by-step updating scheme that readily yields the desired space-time field distribution along the circuit's rim.

Furthermore, the introduced computational procedure has been validated with the help of the eigenfunction expansion method applied to a loss-free rectangular planar circuit, as well as with the aid of FIT applied to an irregularly-shaped planar circuit. TD-CIM may find its applications in TD modeling of various parallel-plane circuit topologies including passive planar circuits, thin microstrip antennas and in solving related signal/power integrity issues in multilayered PCBs. Several extensions and applications of the introduced TD-CIM will be described in detail in the following chapters.

Chapter 4

Relation to the Classic CIM

[1] The FD Contour-Integral Method (CIM) is a well-established numerical technique for the efficient analysis of planar circuits of arbitrary shape [80, 82]. Thanks to its simplicity and versatility, the method is applicable to signal and power integrity analysis of complex high-speed multilayered PCBs, for which it readily provides their parallel-plate impedances [98, 139], as well as to analyzing closely related EMI issues [32, 136].

The CIM formulation is based on a classic 2D contour-integral representation for cylindrical waves (e.g. [82, Eq. (3)]). This integral relation is traditionally solved upon applying the point-matching procedure [45, Sec. 1-4] along with the piecewise-constant field expansion over each of the (relatively small) line segments approximating the circuit's rim. This choice of testing and expansion functions leads to an impedance matrix describing the self- and mutual interactions between the dividing segments, whose elements are easy to calculate [82, Eqs. (14)–(15)]. On the other hand, the point-matching solution may not converge to the actual solution. A straightforward technique that may improve the convergence of the method is to employ the rectangular testing functions. Introducing such a numerical solution is an objective of this chapter.

The following sections are organized as follows. Adopting the problem configuration from the previous Chapter 3, this chapter starts by

---

[1]This chapter is in part based on [132]. The permission from the Radioengineering Journal to reuse the material is gratefully acknowledged.

interrelating the complex-FD reciprocity-based formulation with the classic formulation of CIM as introduced by Okoshi [82]. In Sec. 4.2, the relevant integral equation is solved with the aid of the point-matching procedure. Subsequently in Sec. 4.3, the point-matching solution is generalized upon 'weighting' the integral equation with the sequence of rectangular testing functions. Finally, numerical examples that illustrate convergence properties of the numerical solutions are presented in Sec. 4.4.

4.1   Basic CIM formulation

In order to arrive at the classic CIM formulation, the testing electric-current surface density is chosen to show the Dirac-delta behavior, that is, we substitute $\partial \hat{J}_3^B(\boldsymbol{x}|\boldsymbol{x}^S, s) = \delta(\boldsymbol{x} - \boldsymbol{x}^S)$ in Eq. (3.1) and get

$$
\hat{E}_3(\boldsymbol{x}^S, s) = (s/c\pi) \int_{\boldsymbol{x} \in \partial \Omega} \hat{E}_3(\boldsymbol{x}, s) \mathrm{K}_1 \left[sr(\boldsymbol{x}|\boldsymbol{x}^S)/c \right] \cos[\theta(\boldsymbol{x}|\boldsymbol{x}^S)] \mathrm{d}l(\boldsymbol{x})
$$

$$
+ (s\mu/\pi) \int_{\boldsymbol{x} \in \partial S} \boldsymbol{\nu}(\boldsymbol{x}) \cdot \partial \hat{\boldsymbol{J}}(\boldsymbol{x}, s) \mathrm{K}_0 \left[sr(\boldsymbol{x}|\boldsymbol{x}^S)/c \right] \mathrm{d}l(\boldsymbol{x}) \qquad (4.1)
$$

for $\boldsymbol{x}^S \in \partial \Omega$. Note that this choice of the impulsive testing source is known as the point-matching procedure. Next, taking the limit $\{s = \delta + \mathrm{i}\omega, \delta \downarrow 0, \omega \in \mathbb{R}\}$, the integral equation can be re-written in the real-FD, viz

$$
\hat{V}(\boldsymbol{x}^S, \mathrm{i}\omega) = (k/2\mathrm{i}) \int_{\boldsymbol{x} \in \partial \Omega} \hat{V}(\boldsymbol{x}, \mathrm{i}\omega) \mathrm{H}_1^{(2)} \left[kr(\boldsymbol{x}|\boldsymbol{x}^S) \right] \cos[\theta(\boldsymbol{x}|\boldsymbol{x}^S)] \mathrm{d}l(\boldsymbol{x})
$$

$$
- (\omega \mu d/2) \int_{\boldsymbol{x} \in \partial S} \boldsymbol{\nu}(\boldsymbol{x}) \cdot \partial \hat{\boldsymbol{J}}(\boldsymbol{x}, \mathrm{i}\omega) \mathrm{H}_0^{(2)} \left[kr(\boldsymbol{x}|\boldsymbol{x}^S) \right] \mathrm{d}l(\boldsymbol{x}) \qquad (4.2)
$$

for $\boldsymbol{x}^S \in \partial \Omega$, where $\hat{V} = -d\hat{E}_3$, $k = \omega/c$. In Eq. (4.2) we have used [1, (9.6.4)] to express the modified Bessel functions with the complex argument using the Hankel functions of the second kind.

In the following sections, the integral equation (4.2) is solved numerically. To this end, the circuit's rim is first approximated by a set of line segments $\partial \Omega \simeq \cup_{m=1}^N \Delta \Omega^{[m]}$ (see Fig. 3.1). Subsequently, upon employing the piecewise-constant expansion, the equation is cast into its matrix form, that is

$$
\boldsymbol{U} \cdot \boldsymbol{V} = \boldsymbol{H} \cdot \boldsymbol{I} \qquad (4.3)
$$

in which $\boldsymbol{V}$ is the voltage 1D-array of $[N \times 1]$ (unknown) coefficients, $\boldsymbol{I}$ is the electric-current 1D-array of $[N \times 1]$ (prescribed) excitation coefficients and U and H are $[N \times N]$ 2D-arrays. Clearly, the corresponding $[N \times N]$ impedance matrix directly follows from Eq. (4.3) as $Z = U^{-1} \cdot H$.

4.2 Point-matching solution

In the first step, the unknown voltage distribution along the approximated circuit's periphery is expanded in terms of the rectangular functions (see Eq. (2.44)), that is

$$\hat{V}(\boldsymbol{x}, \mathrm{i}\omega) = \sum_{m=1}^{N} \hat{v}^{[m]} \Pi^{[m]}(\boldsymbol{x}) \tag{4.4}$$

where $\hat{v}^{[m]}$ are the expansion coefficients of vector $\boldsymbol{V}$. Upon enforcing the equality in Eq. (4.2) at isolated points located at the centers of the dividing segments denoted by $\boldsymbol{x}^{[m;c]}$, for all $m = \{1, \cdots, N\}$, we end up with (cf. [82, Sec. III] and [80, Eq. (3.26)])

$$(\boldsymbol{U})_{S,m} = -(k\triangle\Omega^{[m]}/2\mathrm{i})$$
$$\mathrm{H}_1^{(2)}\left[kr(\boldsymbol{x}^{[m;c]}|\boldsymbol{x}^{[S;c]})\right]\cos\left[\theta(\boldsymbol{x}^{[m;c]}|\boldsymbol{x}^{[S;c]})\right] \tag{4.5}$$
$$(\boldsymbol{H})_{S,m} = (\omega\mu d/2)\mathrm{H}_0^{(2)}\left[kr(\boldsymbol{x}^{[m;c]}|\boldsymbol{x}^{[S;c]})\right] \tag{4.6}$$

for all $S \neq m$ and

$$(\boldsymbol{U})_{S,m} = 1 \tag{4.7}$$
$$(\boldsymbol{H})_{S,m} = \frac{\omega\mu d}{2}\left\{1 - \frac{2\mathrm{i}}{\pi}\left[\ln\left(\frac{k\triangle\Omega^{[m]}}{4}\right) - 1 + \gamma\right]\right\} \tag{4.8}$$

for all $S = m$. The latter expression has been found using the small-argument expansion of the Hankel function [1, Eq. (9.1.8)]

$$\mathrm{H}_0^{(2)}(x) = (2\mathrm{i}/\pi)\ln(2/x) - (2\mathrm{i}/\pi)\gamma + 1 + \mathcal{O}(x^2) \text{ as } x \downarrow 0 \tag{4.9}$$

and the following integral (cf. Appendix A)

$$\int_{\lambda=0}^{1} \ln[|(1-\lambda)(-\triangle\Omega/2) + \lambda\triangle\Omega/2|] = \ln(\triangle\Omega/2) - 1 \tag{4.10}$$

for $\{\triangle\Omega \in \mathbb{R}; \triangle\Omega > 0\}$. A sample MATLAB® implementation of the point-matching solution can be found in Sec. C.1.1.

4.3   Pulse-matching solution

Instead of applying the point-matching procedure we can start over with the real-FD version of the reciprocity-based relation (3.1) and associate the corresponding testing-source density with the rectangular function, i.e. we let $\partial \hat{J}_3^B(\boldsymbol{x}|\boldsymbol{x}^S, s) = \Pi^{[S]}(\boldsymbol{x})$. This choice, in combination with the piecewise-constant expansion (4.4), leads to the system of algebraic equations (4.3) with

$$(\boldsymbol{U})_{S,m} = -\left(k\triangle\Omega^{[m]}/2\mathrm{i}\right) \int_{\lambda=0}^{1} \mathrm{d}\lambda \int_{\lambda^T=0}^{1} \mathrm{H}_1^{(2)}\left\{kr[\boldsymbol{x}(\lambda)|\boldsymbol{x}^T(\lambda^T)]\right\}$$
$$\cos\left\{\theta[\boldsymbol{x}(\lambda)|\boldsymbol{x}^T(\lambda^T)]\right\}\mathrm{d}\lambda^T \qquad (4.11)$$

$$(\boldsymbol{H})_{S,m} = (\omega\mu d/2) \int_{\lambda=0}^{1} \mathrm{d}\lambda \int_{\lambda^T=0}^{1} \mathrm{H}_0^{(2)}\left\{kr[\boldsymbol{x}(\lambda)|\boldsymbol{x}^T(\lambda^T)]\right\}\mathrm{d}\lambda^T \quad (4.12)$$

in which

$$\boldsymbol{x}(\lambda) = \boldsymbol{x}^{[m]} + \lambda\left(\boldsymbol{x}^{[m+1]} - \boldsymbol{x}^{[m]}\right) \in \triangle\Omega^{[m]} \qquad (4.13)$$

$$\boldsymbol{x}^T(\lambda^T) = \boldsymbol{x}^{[S]} + \lambda^T\left(\boldsymbol{x}^{[S+1]} - \boldsymbol{x}^{[S]}\right) \in \triangle\Omega^{[S]} \qquad (4.14)$$

for all $S \neq m$. Similarly to the previous section, the diagonal terms are handled analytically. In this way, after a few steps of algebra, we obtain

$$(\boldsymbol{U})_{S,m} = 1 \qquad (4.15)$$

$$(\boldsymbol{H})_{S,m} = \frac{\omega\mu d}{2}\left\{1 - \frac{2\mathrm{i}}{\pi}\left[\ln\left(\frac{k\triangle\Omega^{[m]}}{2}\right) - \frac{3}{2} + \gamma\right]\right\} \qquad (4.16)$$

for all $S = m$. Finally, it is noted that Eqs. (4.5)–(4.6) can be understood as a special case of (4.11)–(4.12) to which the 1-point Gaussian quadrature (see [1, Eq. (25.4.30)]) is applied. An illustrative MATLAB® implementation of the pulse-matching solution can be found in Sec. C.1.2.

4.4   Numerical results

In this section we shall analyse a rectangular planar circuit of dimensions $L = 0.10$ [m] and $W = 0.15$ [m] (see Fig. 4.1). The thickness of the planar circuit is $d = 1.50$ [mm]. The dielectric filling is described by its electric permittivity $\epsilon = 4.50\,\epsilon_0$ and magnetic permeability $\mu = \mu_0$. The corresponding EM wave speed in the dielectric layer is $c = (\epsilon\mu)^{-1/2}$. The circuit is assumed to show low losses such that the corresponding

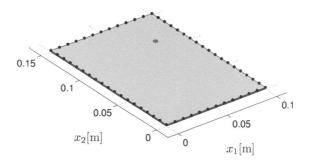

Figure 4.1: Computational model of the analysed rectangular circuit with a vertical excitation port (the dots on the patch).

(complex-valued) wavenumber k can be approximated according to [80, Sec. 2.2.1]

$$k \simeq (\omega/c)\left\{1 + [\tan(\delta) + \delta_s/d]/2\mathrm{i}\right\} \tag{4.17}$$

where the dielectric loss is accounted for via $\tan(\delta) = 0.0045$, the skin depth of the conductor is found from $\delta_s = \sqrt{2/\omega\mu\sigma}$ with conductivity $\sigma = 5.80 \cdot 10^7$ [S/m]. The planar circuit is activated using the excitation vertical port that has its center at $\{x_1^S, x_2^S\} = \{0.075, 0.1125\}$ [m]. The CIM model of the port has a hexagonal cross-section of circum-radius 1.50 [mm]. All the calculations that follow are performed in the frequency range $\{50 \leq f = \omega/2\pi \leq 2000\}$ [MHz] at 200 uniformly-spaced frequency points. The circuit's boundary is discretized such that $\max_n(|\triangle\Omega^{[n]}|) < 0.12\,c/\max(f)$. The integrations in Eqs. (4.11) and (4.12) are carried out using the Gauss-Legendre quadrature, symbolically written as

$$\int_{\lambda=0}^{1} f(\lambda)\mathrm{d}\lambda \simeq \sum_{k=1}^{K} w_k f(\lambda_k) \tag{4.18}$$

where the corresponding abscissas λ_k and weights w_k for $\{0 \leq \lambda \leq 1\}$, $K = \{1,\ldots,8\}$ of the quadrature can be found in [1, p. 921], for example. Recall that for $K = 1$ for which $w_1 = 1$, $\lambda_1 = 1/2$, Eqs. (4.11)–(4.12) become fully equivalent to Eqs. (4.5)–(4.6). For validation purposes, the input impedance is also evaluated using a special case of the double-summation formula (3.18), namely

$$\hat{Z}(\mathrm{i}\omega) = \frac{\mathrm{i}\omega\mu d}{LW} \lim_{M,N\to\infty} \sum_{m=0}^{M} \sum_{n=0}^{N} \frac{e_m^2 e_n^2}{k_m^2 + k_n^2 - k^2} F_{mn}^S F_{mn}^P \tag{4.19}$$

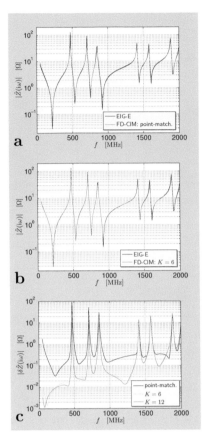

Figure 4.2: Input impedance of the rectangular planar circuit evaluated using FD-CIM and the eigen-function expansion method (EIG-E). (a) Point-matching solution with EIG-E; (b) pulse-matching solution with EIG-E; (c) the absolute error of the FD-CIM solutions with respect to the referential EIG-E solution.

with $F_{mn}^S = F_{mn}^P$, where

$$F_{mn}^S = \cos(k_m x_1^S)\cos(k_n x_2^S)\text{sinc}(k_m W_1^S/2)\text{sinc}(k_n W_2^S/2) \qquad (4.20)$$

where we take $W_1^S = W_2^S = 1.0\,[\text{mm}]$. In the actual calculations, the summations in Eq. (4.19) are truncated to $N = M = 1000$.

The results are summarized in Fig. 4.2. In Figs. 4.2a and 4.2b we have shown the point-matching and pulse-matching solutions, respectively, together with the FD response calculated according to the analytical solution (4.19). As can be observed, both CIM-based solutions correlate well with the reference. In order to clearly assess the accuracy of the numerical solutions, the absolute error of the calculated input impedance with respect to the referential solution (4.19) has been plotted in Fig. 4.2c. Apparently, the calculated error curves attain their peak values at the circuit's resonance frequencies. Comparing the point-matching and pulse-matching approaches, the latter solution leads, except for the very high-frequency part of the frequency range, to more accurate results. The difference is most evident at low frequencies. Finally it has been observed that doubling the number of the integration points from $K = 6$ to $K = 12$ does not lead to a significant improvement. For a related study on the impact of the width of rectangular testing functions on such numerical results' accuracy we refer the reader to [67, Sec. 5.2.2].

4.5 Conclusions

In this chapter, we have demonstrated the link between the reciprocity-based relation given in Chapter 3 and the classic real-FD CIM formulation. The resulting integral equation has been solved for two different types of the testing-source density. In this respect, it has been shown that the classic point-matching solution can be viewed as a special case of the pulse-matching solution to which the 1-point Gaussian quadrature is applied. With the aid of the analytical solution based on the eigenfunction expansion, it has finally been demonstrated that the pulse-matching solution may provide more accurate results than the classic point-matching one. On the other hand, as the pulse-matching solution requires computation of integrals, one should carefully consider whether the improvement is worth the efforts for practical purposes.

Chapter 5

Rectangular Planar Circuits with Relaxation

[1] The emergence of complex high-speed digital interconnecting structures has initiated the research into efficient yet reliable modeling techniques capable of describing pulsed transmission characteristics of planar structures. A group of computational techniques that falls in this category is based on the cavity model that is used in Chapter 3 for the TD analysis of instantaneously-reacting and loss-free planar circuits. Having the indispensable impact of relaxation phenomena on the signal transfer on mind (see e.g. [52]), the main objective of this chapter is to provide an efficient modeling technique that allows us to analyze the space-time field distribution within a rectangular planar circuit with relaxation behavior in its dielectric filling. To that end, the method of images is combined with a robust numerical inversion of the Laplace transformation. In this way, we arrive at the field expansion that can be interpreted as being composed of 'ray-like' TD constituents propagating via reflections against circuit's periphery.

The vast majority of previous works on rectangular power-ground structures tackle the problem traditionally in the real-FD [57–59, 119]. Here, two forms of the fundamental solution, mutually interrelated via the Poisson summation formula [145, Sec. 7.5], can be in principle distinguished. The first form is based on the eigenfunction expansion [57–

---

[1] © [2014] IEEE. This chapter is in part adapted, with permission, from [126].

59], while the second one is represented via the expansion in image sources [119]. Both FD expansions contain infinite summations and their applicability heavily depends on the frequency range of interest [68, Sec. 7.2]. The latter behavior can best be exploited in TD. In particular, it is worth noting that the TD image-source expansion has the property that each 'higher' constituent appears at the field point later then the previous one, which makes it possible to account for only a *finite* number of image terms without any loss of accuracy. This important feature has been previously used in [85], where closed-form expressions for the TD voltage response of rectangular planar structures were introduced. The latter work, however, is limited to loss-free structures and to special cases of the excitation pulse shape. Accordingly, the main purpose of this chapter is the construction of novel space-time expressions describing the pulsed-signal transmission over a rectangular planar circuit showing rather general (Boltzmann-type) relaxation in its dielectric filling [126]. As the obtained closed-form expressions are physically very intuitive and easily implemented they may be readily applied to benchmark purely numerical techniques, for example.

The present chapter is organized as follows. In the first part, the relation between the ray-like solution and the standard eigenfunction expansion is discussed for a rectangular circuit with the instantaneously reacting filling. The second part provides the closed-form solution for a rectangular planar circuit whose dielectric losses are accounted for by its electric conductivity. Subsequently, a general technique that makes it possible to account for relaxation behavior of the dielectric layer is proposed. Potentialities of this technique are demonstrated on Debye's model of an isotropic dielectric. Finally, the obtained results are validated with the aid of FIT.

5.1 Modal and ray-like TD expansions

The initial-boundary value problem defined in Eqs. (2.1)–(2.4) with (2.6) may be solved analytically in closed form for the planar circuit of rectangular shape with $\Omega = \{x \in \mathbb{R}^2; 0 \le x_1 \le L, 0 \le x_2 \le W\}$. In such a case, one may either apply the separation of variable technique [146, Sec. 4.2] or the method of images [146, Sec. 7.5]. The former leads to the classical eigenfunction expansion discussed in Sec. 2.1.1, while the latter yields the ray-type expansion that can be viewed as a collection of rays propagating via reflections against the circuit's boundary. Both approaches are discussed in this section for the impulsive dielectric relaxation function $\kappa(t) = \epsilon \delta(t)$. The incorporation of relaxation effects is addressed in the subsequent sections.

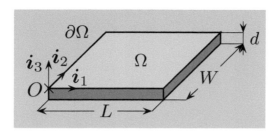

Figure 5.1: Rectangular planar circuit. © [2014] IEEE. Reprinted, with permission, from [126].

The rectangular circuit as shown in Fig. 5.1 is supposed to be activated via a spatially concentrated electric-current source

$$J_3(\boldsymbol{x}, t) = \mathcal{I}(t)\delta(\boldsymbol{x} - \boldsymbol{x}^S) \tag{5.1}$$

with $\mathcal{I}(t) = 0$ for $t < 0$. Considering a vanishing spatial support of the receiving probe, the excited pulsed voltage response observed at $\boldsymbol{x} = \boldsymbol{x}^P$ can be written as

$$\mathcal{V}(t) = \mu d \partial_t \mathcal{I}(t) * G(\boldsymbol{x}^P | \boldsymbol{x}^S, t) \tag{5.2}$$

where we have used the results from Sec. 2.1.1. The complex-FD counterpart of the Green's function satisfies the modified Helmholtz equation (2.9) with the Neumann-type boundary condition (2.10). The boundary-value problem has for the given rectangular shape of Ω the closed-form analytical solution that reads (cf. Eq. (2.16))

$$\hat{G}(\boldsymbol{x} | \boldsymbol{x}^S, s) = \frac{1}{LW} \lim_{M,N \to \infty} \sum_{m=0}^{M} \sum_{n=0}^{N} \frac{e_m^2 e_n^2}{k_m^2 + k_n^2 + \hat{\gamma}^2} F_{mn}(\boldsymbol{x} | \boldsymbol{x}^S) \tag{5.3}$$

with

$$F_{mn} = \cos(k_m x_1) \cos(k_n x_2) \cos(k_m x_1^S) \cos(k_n x_2^S) \tag{5.4}$$

where $\hat{\gamma} = s/c$ for the loss-free, instantaneously reacting dielectric layer, $c = (\epsilon \mu_0)^{-1/2}$ and $e_m = 1$ for $m = 0$ and $e_m = \sqrt{2}$ for $m \neq 0$, $k_m = m\pi/L$, $k_n = n\pi/W$. Equation (5.3) can be with the help of [1, (29.3.15)] transformed into TD, which in combination with (5.2) yields

$$\mathcal{V}(t) = \mu d \partial_t \mathcal{I}(t) * \frac{c}{LW} \sum_{m,n} \frac{e_m^2 e_n^2}{(k_m^2 + k_n^2)^{1/2}}$$

$$F_{mn}(\boldsymbol{x}^P | \boldsymbol{x}^S) \sin \left[ct(k_m^2 + k_n^2)^{1/2} \right] \mathrm{H}(t) \tag{5.5}$$

where we have used the shorthand notation for the double sum in (5.3). The lowest-order term with $m = n = 0$ requires special attention and follows as

$$\mathcal{V}^{[0,0]}(t) = (d/\epsilon LW) \int_{\tau=0}^{t} \mathcal{I}(\tau) d\tau \tag{5.6}$$

which clearly represents the charge accumulation on a parallel-plate capacitor.

An alternative solution relies on the method of images. In this method, the total solution $G(.)$ is composed of the fundamental solution $G_\infty(.)$ satisfying the wave equation with the causality condition and of a secondary part that is adjusted such that $G(.)$ satisfies the (Neumann-type) boundary condition along $\partial\Omega$. In this way we arrive at

$$\mathcal{V}(t) = \mu d\partial_t \mathcal{I}(t) * \sum_{p=-P}^{P} \sum_{q=-Q}^{Q}$$

$$\{G_\infty(x_1^P|x_1^S + 2pL, x_2^P|x_2^S + 2qW, t)|_{++}$$

$$+ G_\infty(x_1^P|x_1^S + 2pL, x_2^P|2qW - x_2^S, t)|_{+-}$$

$$+ G_\infty(x_1^P|2pL - x_1^S, x_2^P|x_2^S + 2qW, t)|_{-+}$$

$$+ G_\infty(x_1^P|2pL - x_1^S, x_2^P|2qW - x_2^S, t)|_{--}\} \tag{5.7}$$

where

$$G_\infty(x_1^P|x_1^S, x_2^P|x_2^S, t) = (1/2\pi)(t^2 - r^2/c^2)^{-1/2}\mathrm{H}(t - r/c) \tag{5.8}$$

is the lossless two-dimensional fundamental solution and $r = r(\boldsymbol{x}^P|\boldsymbol{x}^S)$ denotes the Eucledian distance between the source and field points, that is

$$r(\boldsymbol{x}^P|\boldsymbol{x}^S) = \left[(x_1^P - x_1^S)^2 + (x_2^P - x_2^S)^2\right]^{1/2} \tag{5.9}$$

Upon inspection of (5.7) with (5.8) we may identify the arrival times of the corresponding TD constituents, that is

$$c\,T_{++}^{[pq]} = \left[(x_1^P - x_1^S - 2pL)^2 + (x_2^P - x_2^S - 2qW)^2\right]^{1/2} \tag{5.10}$$

$$c\,T_{+-}^{[pq]} = \left[(x_1^P - x_1^S - 2pL)^2 + (x_2^P + x_2^S - 2qW)^2\right]^{1/2} \tag{5.11}$$

$$c\,T_{-+}^{[pq]} = \left[(x_1^P + x_1^S - 2pL)^2 + (x_2^P - x_2^S - 2qW)^2\right]^{1/2} \tag{5.12}$$

$$c\,T_{--}^{[pq]} = \left[(x_1^P + x_1^S - 2pL)^2 + (x_2^P + x_2^S - 2qW)^2\right]^{1/2} \tag{5.13}$$

From Eqs. (5.10)–(5.13) it is immediately clear that the arrival times increase with $|p|+|q|$, which is proportional to the number of reflections against the circuit's edges. Consequently, the TD constituents in (5.7) arrive at the field point in a successive manner and, hence, in any finite time window of observation, only a finite number of them is necessary to construct the exact solution. On the other hand, the situation is very different for the modal expansion (5.5). In it, all the modal constituents start at the origin, $t = 0$, and to obtain the exact solution one would need to include an unlimited number of them. In practice, of course, their number is always truncated once the prescribed precision is reached. Further properties of the modal and ray-like expansions will be discussed in Sec. (5.4). For related works on the subject we refer the reader to [124, 137].

5.2 Conduction-loss dielectric relaxation

A straightforward way to model a lossy dielectric compatible with the property of causality is to specify its relative permittivity ϵ_r and electric conductivity σ. The corresponding dielectric relaxation function, therefore, has the following form

$$\kappa(t) = \epsilon_0[\epsilon_r \delta(t) + (\sigma/\epsilon_0)H(t)] \tag{5.14}$$

Presence of non-zero conducting current in (2.1) manifests itself by a diffusive term in the corresponding dissipative wave equation

$$(\partial_1^2 + \partial_2^2)G_\infty - c^{-2}(\partial_t^2 + \tau_c^{-1}\partial_t)G_\infty = -\delta(\boldsymbol{x} - \boldsymbol{x}^S)\delta(t) \tag{5.15}$$

that is next solved for all $\boldsymbol{x} \in \mathbb{R}^2$ and all $t > 0$ together with the zero initial conditions and the condition of causality. Here, $\tau_c = \epsilon/\sigma$ is the conduction relaxation time and $c = (\epsilon_r\epsilon_0\mu_0)^{-1/2}$ is the corresponding EM wave speed. To solve Eq. (5.15) analytically, one may either apply the extended Cagniard-DeHoop method [25] or start with the solution of the three-dimensional dissipative wave equation [24, Sec. 26.5] and apply Hadamard's method of descent [17, III -§4.4]. The latter procedure was used in [68, Sec. 7.4], for example. Based on the results given in Appendix H, the fundamental solution is written as (cf. Eq. (H.12))

$$G_\infty(x_1^P|x_1^S, x_2^P|x_2^S, t) = (1/2\pi)(t^2 - r^2/c^2)^{-1/2}H(t - r/c)$$
$$\left\{1 + 2\sinh^2\left[(t^2 - r^2/c^2)^{1/2}/4\tau_c\right]\right\}\exp(-t/2\tau_c) \tag{5.16}$$

Evidently Eq. (5.8) is a special case of (5.16) for vanishing conductivity $\sigma \downarrow 0$. Its first part represents an attenuated fundamental solution of the loss-free wave equation, while the second part represents a dispersive contribution introduced by the diffusive term in (5.15). Finally, the total electric field follows upon substituting of (5.16) in (5.7).

5.3 Debye's dielectric relaxation

The relaxation behavior of an isotropic dielectric slab may be modeled via the first-order Debye model (see [48, Sec. 8.4], for instance). The corresponding dielectric relaxation function has the following form

$$\kappa(t) = \epsilon_0\{\epsilon_\infty\delta(t) + [(\epsilon_r - \epsilon_\infty)/\tau_r]\exp(-t/\tau_r)H(t)\} \tag{5.17}$$

where ϵ_r and ϵ_∞ are the characteristic relative permittivities for which $0 < \epsilon_\infty < \epsilon_r$ and τ_r is the relaxation time. The corresponding wave equation then reads (cf. Eq. (J.1))

$$(\partial_1^2 + \partial_2^2)G_\infty - c_\infty^{-2}\partial_t^2\Big\{G_\infty + (\epsilon_r/\epsilon_\infty - 1)\tau_r^{-1}$$

$$\int_{\tau=-\infty}^{t} \exp[-(t-\tau)/\tau_r]G_\infty(\boldsymbol{x},\tau)\mathrm{d}\tau\Big\} = -\delta(\boldsymbol{x} - \boldsymbol{x}^S)\delta(t) \tag{5.18}$$

where $c_\infty = (\epsilon_\infty\epsilon_0\mu_0)^{-1/2}$. Following the method described in Appendix J, the fundamental solution is expressed as

$$G_\infty(x_1|x_1^S, x_2|x_2^S, t) = \frac{1}{2\pi}\frac{H(t - r/c_\infty)}{(t^2 - r^2/c_\infty^2)^{1/2}}\exp\left[-(t/2\tau_r)(\epsilon_r - \epsilon_\infty)\right]$$

$$+ \frac{1}{2\pi}\int_{\tau=r/c_\infty}^{t}\frac{F(t,\tau)\,\mathrm{d}\tau}{(\tau^2 - r^2/c_\infty^2)^{1/2}} \tag{5.19}$$

for $t > \tau$, where the integrated function $F(t,\tau)$ follows from the Bromwich integral

$$F(t,\tau) = \frac{1}{2\pi\mathrm{i}}\int_{s\in\mathcal{B}}\exp(st)\hat{F}(s,\tau)\mathrm{d}s \tag{5.20}$$

where

$$\hat{F}(s,\tau) = \exp\left[-\hat{L}(s)\tau\right] - \exp\left[-\hat{L}_\infty(s)\tau\right] \tag{5.21}$$

$$\hat{L}(s) = s\left[(s+\alpha)/(s+\beta)\right]^{1/2} \tag{5.22}$$

$$\hat{L}_\infty(s) = s + (\alpha - \beta)/2 \tag{5.23}$$

with $\alpha = (\epsilon_r/\epsilon_\infty)/\tau_r$ and $\beta = 1/\tau_r$. Here, $\hat{L}_\infty(s)$ denotes the leading terms of $\hat{L}(s)$ in the asymptotic expansion as $|s| \to \infty$. The Bromwich integration contour $\mathcal{B}$ in (5.20) runs parallel to $\mathrm{Re}(s) = 0$ and is shifted to the right of all singularities in the complex s-plane. Here, we encounter two algebraic branch points on the negative real axis at $s = \{-\alpha, -\beta\}$. The corresponding branch cuts are chosen such that $\mathrm{Re}[(s + \alpha)^{1/2}] \geq 0$ and $\mathrm{Re}[(s + \beta)^{1/2}] \geq 0$ for all $s \in \mathbb{C}$, which implies two overlapping branch cuts along the negative real axis $\{s \in \mathbb{C}; -\infty < \mathrm{Re}(s) \leq -\alpha, \mathrm{Im}(s) = 0\}$ and $\{s \in \mathbb{C}; -\infty < \mathrm{Re}(s) \leq -\beta, \mathrm{Im}(s) = 0\}$. The Bromwich contour is then in virtue of Jordan's lemma closed to the right and the resulting contour is, in view of Cauchy's theorem, contracted to a new contour $\Gamma \cup \Gamma^*$ along which the integral is carried out numerically. For details concerning the hyperbolic contour we refer the reader to Appendix I that is further supplemented with a demo implementation and an illustrative numerical example.

Similarly to the previous section, the fundamental solution (5.19) consists of two parts. The first one is an attenuated two-dimensional Green's function of the corresponding loss-free wave equation and the second one represents the relaxation behavior. The latter part vanishes close to the wavefront at $t = r/c_\infty$, which is, in fact, a typical feature of dispersive phenomena [37].

5.4 Numerical results

In order to provide an application of the results introduced in the previous sections, sample calculations are performed for a rectangular power-ground structure of dimensions $L = 100 \, [\mathrm{mm}]$ and $W = 75.0 \, [\mathrm{mm}]$ (see Fig. 5.1). The rectangular circuit is excited by the triangular electric-current pulse (see Fig. 5.2 and Eq. (F.4))

$$\mathcal{I}(t) = 2A \left[\frac{t}{t_{\mathrm{w}}} \mathrm{H}(t) - 2 \left(\frac{t}{t_{\mathrm{w}}} - \frac{1}{2} \right) \mathrm{H} \left(\frac{t}{t_{\mathrm{w}}} - \frac{1}{2} \right) \right.$$

$$\left. + \left(\frac{t}{t_{\mathrm{w}}} - 1 \right) \mathrm{H} \left(\frac{t}{t_{\mathrm{w}}} - 1 \right) \right] \qquad (5.24)$$

via a spatially localized vertical port placed at $\{x_1^S, x_2^S\} = \{25.0, 20.0\} \, [\mathrm{mm}]$ (see Fig. 5.3). Here, A is the pulse amplitude, t_{w} corresponds to the length of the base and $t_{\mathrm{w}}/2$ is equal to the pulse rise and fall time. For the following examples we choose $A = 1.0 \, [\mathrm{A}]$ and $v \, t_{\mathrm{w}} = 0.10 \, [\mathrm{m}]$. The pulse voltage response is observed at $\{x_1, x_2\} = \{80.0, 60.0\} \, [\mathrm{mm}]$ (see Fig. 5.3) within the finite time window of observation $\{0 \leq vt \leq 1.0\} \, [\mathrm{m}]$. Here, $v = c = (\epsilon_{\mathrm{r}} \epsilon_0 \mu_0)^{-1/2}$ or $v = c_\infty =$

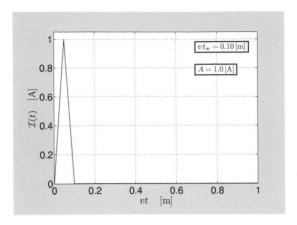

Figure 5.2: The triangular electric-current excitation signature. © [2014] IEEE. Adapted, with permission, from [126].

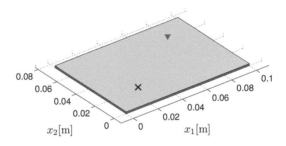

Figure 5.3: Model of the analysed rectangular planar circuit with a field probe (the solid triangle) and an excitation port (the cross symbol).

$(\epsilon_\infty \epsilon_0 \mu_0)^{-1/2}$ for the conduction-loss or Debye dielectric relaxation, respectively.

5.4.1 Modal and ray time-domain constituents

The first example is related to Sec. 5.1 and demonstrates the main features of the modal and ray time-domain expansions. The EM properties of the dielectric layer are described here by its scalar electric permittivity $\epsilon = 4.50\,\epsilon_0$ and magnetic permeability μ_0. Four low-order terms from the right-hand side of Eq. (5.5) are shown in Fig. 5.4a.

As already noted in Sec. 5.1, all the modal constituents start at $t = 0$ and the hint of causality starts to appear after a sufficient number of them is included. For the field evaluation itself it is important to observe that as the rate of oscillations increases with the order of the

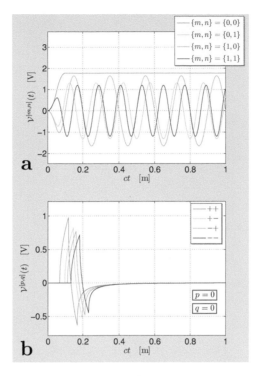

Figure 5.4: Time-domain constituents. (a) The oscillatory modal type; (b) the progressing ray type. © [2014] IEEE. Reprinted, with permission, from [126].

constituents, which may cause difficulties in the numerical calculation of the time convolution with (the time derivative of) the excitation pulse shape for (strongly oscillatory) high-order terms. This is not an issue for special cases of the excitation pulse shape, such as the triangular one (5.24), for instance, for which the time convolution integral can be evaluated in closed form. The pulse shapes of low-order ray constituents, as calculated using Eq. (5.7), are shown in Fig. 5.4b. Here we may observe the direct source/probe wave constituent $(++)$ as well as the reflected constituents that are associated with the image sources. As can be seen in Fig. 5.4b, all the reflected constituents reach the field point after the arrival of the direct wave, which makes the evaluated signal strictly causal. This fact is illustrated in Fig. 5.5a, where the total field responses, as evaluated according to Eq. (5.5) with $M = N = 25$ and Eq. (5.7) with $P = Q = 7$, are compared. While the ray-type expansion already provides the exact results, the modal solution is apparently still missing 'high-frequency components' (see Fig. 5.5b). In this example, the total number of the modal constituents is $M \times N = 625$, while the

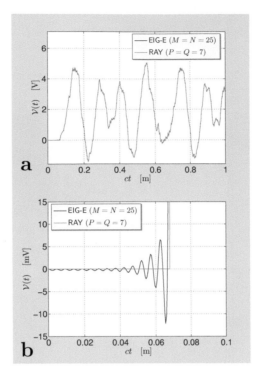

Figure 5.5: The voltage responses represented using (a) the ray-like (RAY) and modal solutions (EIG-E); (b) the early-time part of the responses.

total number of the ray constituents that appear in the chosen time window is only 423 out of $4 \times (2P + 1) \times (2Q + 1) = 900$.

5.4.2   Inclusion of conduction loss

The second example illustrates the results discussed in Sec. 5.2. The EM properties of the dielectric slab are now described with its scalar electric permittivity $\epsilon = 4.50 \, \epsilon_0$, electric conductivity $\sigma = 0.02 \, [\text{S/m}]$ and magnetic permeability μ_0. The corresponding conduction relaxation time τ_c is a small fraction of the excitation pulse time width $t_w/2$, namely $t_w/\tau_c \simeq 3 \cdot 10^4$.

The following examples were evaluated through the use of Eqs. (5.7) with (5.16) as well as with FIT as implemented in CST Microwave Studio®. The reference FIT-based computational model of the analysed structure is placed in the homogeneous, isotropic and loss-free ('normal') embedding. This model consists of a homogeneous layer described with the corresponding electric permittivity, electric conductivity, magnetic permeability and thickness $d = 1.50 \, [\text{mm}]$. Note that the

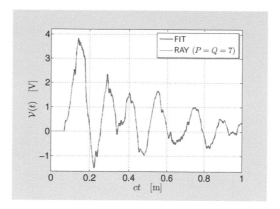

Figure 5.6: The pulsed voltage responses of the circuit with the conductive-loss dielectric relaxation as evaluated using the ray-type expansion (RAY) and the referential FIT.

layer is very thin with respect to the spatial support of the excitation pulse, $d/ct_w = 0.015$. The layer is placed in between two PEC sheets of vanishing thickness. The model is finely discretized into about 600 thousand hexahedral mesh cells. The sidewalls of the surrounding box are defined as magnetic walls with the vanishing tangential magnetic-field components. The structure is excited through the (electric-current-type) discrete port having a vanishing radius. The resulting voltage signals are shown in Fig. 5.6. As can be seen, the corresponding results correlate very well. In fact, the ray-type expansion is 'exact' and thus provides an useful tool for benchmarking purely numerical techniques such as FIT.

5.4.3   *Inclusion of Debye's dielectric relaxation*

The last example provides illustrative results concerning the Debye relaxation function, as discussed in Sec. 5.3. The following pulsed responses were evaluated with the help of Eqs. (5.7), (5.19) and again, with FIT as implemented in CST Microwave Studio®. In the first step, the dispersion characteristics of 'FR-4 (lossy)' as defined in the CST Material Library ('CST MLib') were used to find the parameters of the corresponding dielectric relaxation function (5.17) ('model'). The corresponding complex (steady-state) dielectric relaxation function reads

$$\mathrm{Re}[\hat{\kappa}(\mathrm{i}\omega)]/\epsilon_0 = \epsilon_\infty + (\epsilon_r - \epsilon_\infty)/(1 + \omega^2\tau^2) \tag{5.25}$$

$$\mathrm{Im}[\hat{\kappa}(\mathrm{i}\omega)]/\epsilon_0 = (\epsilon_r - \epsilon_\infty)\omega\tau/(1 + \omega^2\tau^2) \tag{5.26}$$

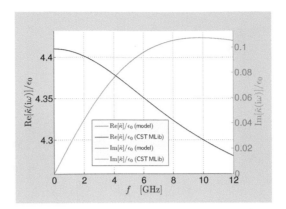

Figure 5.7: Real (Re) and imaginary (Im) parts of the complex dielectric relaxation function related to FR-4. The curves from the implemented model and CST Material Library overlap each other. © [2014] IEEE. Adapted, with permission, from [126].

for $\omega = 2\pi f \in \mathbb{R}$. Its real and imaginary parts for $\epsilon_r = 4.410$, $\epsilon_\infty = 4.195$ and $\tau_r = 1.630 \cdot 10^{-11}$ [s] are plotted in Fig. 5.7. In the real-frequency range $f = (0...12)$ [GHz], that covers the four main lobes of the amplitude-frequency spectrum of the excitation pulse, the dispersion characteristics are almost identical. Except for the dielectric filling, the FIT model was defined as described in the previous subsection 5.4.2. Figure 5.8 shows the corresponding pulsed voltage responses. Again, the calculated pulse shapes correlate very well, thereby validating the proposed methodology.

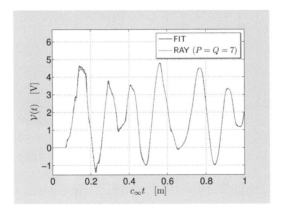

Figure 5.8: The pulsed voltage responses of the circuit with the Debye-dielectric relaxation as evaluated using the ray-type expansion (**RAY**) and the referential FIT.

5.5 Conclusions

An efficient modeling technique for analyzing the pulsed EM-field transmission over a rectangular planar circuit has been described. This technique is based on the method of images and allows for the inclusion of rather general relaxation mechanisms. The TD analysis has been carried out first for the loss-free case, demonstrating the main features of the ray-type and modal expansions. Subsequently, two dielectric relaxation models have been analysed in detail. Namely, attention has been paid to the conduction-loss and Debye-dielectric relaxation functions for an isotropic dielectric.

The proposed technique can be easily implemented, it is physically intuitive and computationally efficient. Furthermore, its combination with the general description of the 2D wave equation with relaxation (see Appendix J) allows us to analyze the impact of complex (Boltzmann-type) relaxation behavior on the signal-transmission characteristics of rectangular planar structures. All these properties make the introduced computational model suitable for the fast broadband EM analysis of rectangular circuits, as well as for benchmarking computational techniques such as the TD finite-difference method, for instance.

The ray-type fundamental solution can be directly applied to the analysis of TD EM radiated emissions and susceptibility of rectangular planar circuits, an example of which is given in Chapter 12. Furthermore, the high efficiency of this TD solution is also employed in Chapter 15, where a compensation-based TD IE technique is formulated and successfully validated.

Chapter 6

Arbitrarily-shaped Planar Circuits with Radiation Loss and Relaxation

[1] The two-dimensional circuit model with the ideal open-circuit boundary represents a loss-free resonator whose TD response never subsides. In order to set up a more realistic computational model, one can account for relaxation and dissipation effects of the dielectric layer and radiation loss. The inclusion of these phenomena in the TD-CIM-based methodology is the main purpose of this chapter.

In general, the dissipation and relaxation mechanisms in the FD description manifest themselves in the imaginary part of the wavenumber, making their inclusion straightforward [80, Sec. 2.2.1]. As far as the TD analysis is concerned, their inclusion often presents a challenging task (see e.g. [40]). In TD, the inclusion of material dissipation and relaxation phenomena changes the form of the field equations and hence the corresponding fundamental solution (see Appendix H). Then, except for special cases, the time convolutions occurring in the reciprocity relations must be calculated numerically and the analysis becomes computationally more involved.

Referring to the cavity model, the effects of non-perfectly conducting planes can be neglected, due to a range of practical problems (see [121, Fig. 8]). This is not the case, however, for its radiation loss. The

---

[1] © [2015] IEEE. This chapter is in part adapted, with permission, from [129].

latter is frequently accounted for by increasing the actual value of the dielectric loss tangent, thereby describing the total loss effect via the effective (or modified) loss tangent [43, Sec. 9.2.2], [4, Sec. 14.2.2]. While this approach has been successfully applied in the real-FD, it can hardly lead to plausible outcomes in TD. Accordingly, radiation losses are addressed here by formulating an admittance-wall boundary condition through which the planar structure is coupled to its exterior embedding. Moreover, two types of dielectric relaxation functions describing material losses in the dielectric layer are discussed and implemented into TD-CIM. In this regard, it is demonstrated that a promising way to solve the aforementioned difficulties is the numerical Laplace-transform inversion based on the deformation of the Bromwich contour into a hyperbolic one (see Appendix I). Although we primarily focus here on the relaxation behavior of an isotropic dielectric described via the finite-conductivity and Debye-type relaxation models, the proposed methodology (see [129]) is very general and allows for the inclusion of more complex (Boltzmann-type) relaxation characteristics such the Lorentzian absorption line, for instance. Finally, a few sample calculations are performed and their results are validated using FIT.

6.1 Formulation of the admittance-wall condition

The point of departure for our analysis is, once again, the reciprocity relation (2.34), namely, the second term on its right-hand side, that is

$$\int_{\boldsymbol{x}\in\partial\Omega} \hat{E}_3^B(\boldsymbol{x}|\boldsymbol{x}^S, s)\boldsymbol{\nu}(\boldsymbol{x}) \cdot \partial\hat{\boldsymbol{J}}(\boldsymbol{x}, s)\mathrm{d}l(\boldsymbol{x}) \qquad (6.1)$$

Now, instead of applying the perfect magnetic-wall boundary condition (2.6) let us impose, along the source-free portion of the circuit's rim, the admittance-wall boundary condition

$$(\boldsymbol{i}_3 \times \boldsymbol{\nu}) \cdot \boldsymbol{H}(\boldsymbol{x} + \delta\boldsymbol{\nu}, t) = -\eta_{\mathrm{w}}(\boldsymbol{x}, t) * E_3(\boldsymbol{x} + \delta\boldsymbol{\nu}, t) \quad \text{as } \delta \downarrow 0 \quad (6.2)$$

for all $\boldsymbol{x} \in \partial\Omega/\partial\mathcal{S}$ and $t > 0$. The boundary condition (6.2) relates the tangential component of the magnetic field strength with the (unknown) electric-field strength through the (causal) plane-wave wall admittance η_{w}. Note that the perfect magnetic-wall boundary condition (2.6) is in fact a special case of the admittance-wall condition (6.2) for which η_{w} and, hence, the tangential component of the magnetic-field strength is zero along the source-free boundary. Upon combining Eq. (6.1) with Eq. (2.35) together with the s-domain counterpart of the

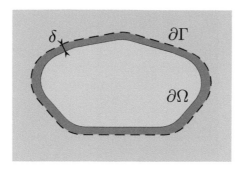

Figure 6.1: Model of the admittance-wall radiating boundary. © [2017] IEEE. Adapted, with permission, from [132].

admittance-wall condition (6.2), we end up with

$$
s\mu \int_{\boldsymbol{x}^T \in \partial\Omega} \partial\hat{J}_3^B(\boldsymbol{x}^T|\boldsymbol{x}^S, s)
$$

$$
\int_{\boldsymbol{x} \in \partial\Gamma} \hat{\eta}_w(\boldsymbol{x}, s)\hat{E}_3(\boldsymbol{x}, s)\hat{G}_\infty[r(\boldsymbol{x}|\boldsymbol{x}^T), s]\mathrm{d}l(\boldsymbol{x})\mathrm{d}l(\boldsymbol{x}^T) \quad (6.3)
$$

where $\partial\Gamma$ denotes the source-free part of the circuit periphery that in line with (6.2) is placed at a distance δ away from the actual circuit's rim (see Fig. 6.1).

Let us next consider, for the sake of simplicity, the loss-free Green's function along with the uniform and instantaneously reacting TD admittance $\eta_w(\boldsymbol{x}, t) = \eta_0\delta(t)$. Then, along the lines described in Chapter 3, the discretized form of Eq. (2.34) can be written as

$$
\left(\boldsymbol{I} - \boldsymbol{Q}_{[0]} + \boldsymbol{W}_{[0]}\right) \cdot \boldsymbol{E}_{[p]} = \sum_{k=1}^{p-1} \left(\boldsymbol{Q}_{[p-k]} - \boldsymbol{W}_{[p-k]}\right) \cdot \boldsymbol{E}_{[k]} + \boldsymbol{F}_{[p]} \quad (6.4)
$$

for all $p = \{1, \cdots, NT\}$. With reference to Eq. (3.8), it is clear that the effect of losses is included via the 3D-array $\boldsymbol{W}$ of dimensions $[N \times N \times NT]$. The elements of this array can be expressed as

$$
(\boldsymbol{W}_{[p-k]})_{S,m} = \frac{1}{\pi c \triangle t}\frac{\eta_0}{\eta}\int_{\boldsymbol{x}^T \in \partial\Omega} T^{[S]}(\boldsymbol{x}^T)\int_{\boldsymbol{x} \in \partial\Gamma} T^{[m]}(\boldsymbol{x})
$$

$$
\Theta[r(\boldsymbol{x}|\boldsymbol{x}^T), (p-k)\triangle t]\mathrm{d}l(\boldsymbol{x}^T)\mathrm{d}l(\boldsymbol{x}) \quad (6.5)
$$

where $\eta = (\epsilon/\mu)^{1/2}$ and

$$
\Theta(r, t) = \zeta(r, t + \triangle t) - 2\zeta(r, t) + \zeta(r, t - \triangle t) \quad (6.6)
$$

$$
\zeta(r, t) = \ln\left[ct/r + \left(c^2t^2/r^2 - 1\right)^{1/2}\right]\mathrm{H}(t - r/c) \quad (6.7)
$$

Here, owing to the spatial shift δ of the radiating boundary, the handling of the self-coupling terms is not necessary. This shift is, in numerical implementations, taken as being a small fraction of the spatial support of the excitation electric-current pulse. The corresponding elements of the tridiagonal 2D-array $\boldsymbol{I}$ are given in Eq. (3.9) and the space-time functions $\psi(r,t)$ and $\Psi(r,t)$ that constitute the elements of $\boldsymbol{Q}_{[p-k]}$ are specified in Eqs. (3.12) and (3.13). The self-coupling elements of $\boldsymbol{Q}_{[p-k]}$ yield the zero contribution as $\boldsymbol{x}$ approaches $\boldsymbol{x}^T$ (see Appendix A).

More elaborate admittances $\eta_{\mathrm{w}} = \eta_{\mathrm{w}}(\boldsymbol{x},s)$ accounting for the edge effects [43, Sec. 9.4] can be included in TD-CIM along the same lines. For instance, we may write

$$(\boldsymbol{W}_{[p-k]})_{S,m} = \frac{1}{\pi}\frac{\mu_0}{L_0}\int_{\boldsymbol{x}^T\in\partial\Omega}T^{[S]}(\boldsymbol{x}^T)\int_{\boldsymbol{x}\in\partial\Gamma}T^{[m]}(\boldsymbol{x})$$
$$\Theta[r(\boldsymbol{x}|\boldsymbol{x}^T),(p-k)\triangle t]\mathrm{d}l(\boldsymbol{x}^T)\mathrm{d}l(\boldsymbol{x}) \qquad (6.8)$$

with

$$\Theta(r,t) = \zeta(r,t+\triangle t) - \zeta(r,t) \qquad (6.9)$$

applying to a purely inductive admittance wall described by $\hat{\eta}_{\mathrm{w}}(\boldsymbol{x},s) = 1/sL_0$ and

$$(\boldsymbol{W}_{[p-k]})_{S,m} = \frac{1}{\pi(c\triangle t)^2}\frac{C_0}{\epsilon}\int_{\boldsymbol{x}^T\in\partial\Omega}T^{[S]}(\boldsymbol{x}^T)\int_{\boldsymbol{x}\in\partial\Gamma}T^{[m]}(\boldsymbol{x})$$
$$\Theta[r(\boldsymbol{x}|\boldsymbol{x}^T),(p-k)\triangle t]\mathrm{d}l(\boldsymbol{x}^T)\mathrm{d}l(\boldsymbol{x}) \quad (6.10)$$

with

$$\Theta(r,t) = \zeta(r,t+\triangle t) - 3\zeta(r,t)$$
$$+ 3\zeta(r,t-\triangle t) - \zeta(r,t-2\triangle t) \qquad (6.11)$$

applying to a purely capacitive admittance wall described by $\hat{\eta}_{\mathrm{w}}(\boldsymbol{x},s) = sC_0$. A demo MATLAB® implementation of the resistive boundary condition can be found in Appendix D. For an alternative numerical example we refer the reader to [132].

6.2 Inclusion of relaxation behavior

We shall analyse a planar circuit excited via the (controlled) vertical electric-current surface density $\partial\hat{J}_3$ applied to a part of the rim $\partial\mathcal{S} \subset \partial\Omega$. The rest of the circuit periphery forms the perfect magnetic wall to which the explicit-type boundary condition (2.6) applies.

The corresponding reciprocity-integral relation has the following form (cf. Eq. (3.1))

$$\int_{\boldsymbol{x}\in\partial\Omega} \hat{E}_3(\boldsymbol{x},s)\partial\hat{J}_3^B(\boldsymbol{x}|\boldsymbol{x}^S,s)\mathrm{d}l(\boldsymbol{x})$$

$$= [\hat{\gamma}(s)/\pi]\int_{\boldsymbol{x}\in\partial\Omega} \hat{E}_3(\boldsymbol{x},s)\int_{\boldsymbol{x}^T\in\partial\Omega} \mathrm{K}_1\left[\hat{\gamma}(s)r(\boldsymbol{x}|\boldsymbol{x}^T)\right]$$
$$\partial\hat{J}_3^B(\boldsymbol{x}^T|\boldsymbol{x}^S,s)\cos[\theta(\boldsymbol{x}|\boldsymbol{x}^T)]\mathrm{d}l(\boldsymbol{x}^T)\mathrm{d}l(\boldsymbol{x})$$

$$+(s\mu/\pi)\int_{\boldsymbol{x}\in\partial S}\partial\hat{J}_3(\boldsymbol{x},s)\int_{\boldsymbol{x}^T\in\partial\Omega} \mathrm{K}_0\left[\hat{\gamma}(s)r(\boldsymbol{x}|\boldsymbol{x}^T)/c\right]$$
$$\partial\hat{J}_3^B(\boldsymbol{x}^T|\boldsymbol{x}^S,s)\mathrm{d}l(\boldsymbol{x}^T)\mathrm{d}l(\boldsymbol{x}) \tag{6.12}$$

where $\hat{\gamma} = \hat{\gamma}(s)$ is the propagation coefficient (see Sec. 2.1.1) that depends on relaxation behavior of the dielectric slab. Besides the instantaneously reacting medium for which $\hat{\gamma}(s) = s/c$ (see Chapter 3), two other models of an isotropic dielectric are analysed, namely:

∎ The finite-conductivity relaxation model, defined via its relative electric permittivity ϵ_r and electric conductivity σ. The corresponding propagation coefficient has the form $\hat{\gamma}(s) = [s(s + \alpha)]^{1/2}/c$ with $c = (\epsilon_r\epsilon_0\mu_0)^{-1/2}$, where α is related to the conduction relaxation time as $\alpha = 1/\tau_c = \sigma/\epsilon$.

∎ The Debye dielectric relaxation model, defined via its relative characteristic permittivities ϵ_r and ϵ_∞ and the relaxation time τ_r. The corresponding propagation coefficient has the form $\hat{\gamma}(s) = s[(s + \alpha)/(s + \beta)]^{1/2}/c_\infty$ with $c_\infty = (\epsilon_\infty\epsilon_0\mu_0)^{-1/2}$, where $\alpha = (\epsilon_r/\epsilon_\infty)/\tau_r$, $\beta = 1/\tau_r$.

Details concerning the Laplace inversion are discussed for both cases in the following subsections.

6.2.1 Conduction-loss dielectric model

The following analysis is carried out for the piecewise linear temporal expansion of the electric field strength (see Chapter 3). Other types of time expansion may be readily handled in a similar way. With reference to Eq. (6.12), the inverse Laplace transformation of the following expressions will be found

$$\hat{\gamma}(s)\mathrm{K}_1[\hat{\gamma}(s)r]/s^2 = [s(s+\alpha)]^{1/2}\mathrm{K}_1\left\{[s(s+\alpha)]^{1/2}r/c\right\}/cs^2 \tag{6.13}$$

$$\mathrm{K}_0[\hat{\gamma}(s)r] = \mathrm{K}_0\left\{[s(s+\alpha)]^{1/2}r/c\right\} \tag{6.14}$$

where $\alpha = \sigma/\epsilon$, $\epsilon = \epsilon_r\epsilon_0$ and $c = (\epsilon_r\epsilon_0\mu_0)^{-1/2}$. It is worth observing that in the limit with $\sigma \downarrow 0$, one may find the corresponding Laplace inversions analytically with the aid of (cf. Eqs. (3.13) and (3.14))

$$\mathcal{L}^{-1}[K_1(sr/c)/s] = (c^2t^2/r^2 - 1)^{1/2}H(t - r/c) \tag{6.15}$$

$$\mathcal{L}^{-1}[K_0(sr/c)] = (t^2 - r^2/c^2)^{-1/2}H(t - r/c) \tag{6.16}$$

Although the inversion of expressions (6.13) and (6.14) can be in principle carried out analytically with the help of the Schouten-van-der-Pol theorem [104, 120], the numerical procedure described in Appendix I is, for the sake of generality, applied here. To that end, one needs to study analytical properties of the expressions in the complex s-plane. Here, we encounter two branch points on the negative real axis at $s = \{0, -\alpha\}$ and a (double) pole singularity at $s = 0$ in Eq. (6.13). The corresponding branch cuts are chosen such that $\mathrm{Re}[(s + \alpha)^{1/2}] \geq 0$ and $\mathrm{Re}(s^{1/2}) \geq 0$ for all $s \in \mathbb{C}$, which implies two overlapping branch cuts along the negative real axis $\{s \in \mathbb{C}; -\infty < \mathrm{Re}(s) \leq 0, \mathrm{Im}(s) = 0\}$, $\{s \in \mathbb{C}; -\infty < \mathrm{Re}(s) \leq -\alpha, \mathrm{Im}(s) = 0\}$ together with

$$c\hat{\gamma}(s) = s + \alpha + \mathcal{O}(s^{-1}) \tag{6.17}$$

as $|s| \to \infty$. On account of the large-argument expansions

$$K_{0,1}(sr/c) = (\pi c/2r)^{1/2}s^{-1/2}\exp(-sr/c)\left[1 + \mathcal{O}(s^{-1})\right] \tag{6.18}$$

as $|s| \to \infty$, the original Bromwich integration contour can be closed to the right for $t > r/c$ and the corresponding integration is in view of Cauchy's theorem carried out along the hyperbolic contour $\Gamma \cup \Gamma^*$ (here $*$ denotes complex conjugate) sketched in Fig. 6.2. Here, only the non-overlapping part of the branch cuts is depicted. For further details concerning the numerical integration along the hyperbolic contour we refer the reader to Appendix I.

Finally, the unknown expansion coefficients follow upon solving Eq. (3.8), provided that the integrated functions ψ and Φ defined in Eqs. (3.13) and (3.14) are replaced by their corresponding (numerical) inversions

$$\psi(r,t) = \mathcal{L}^{-1}\left\{[s(s+\alpha)]^{1/2}K_1\left\{[s(s+\alpha)]^{1/2}r/c\right\}/s^2\right\} \tag{6.19}$$

$$\Phi(r,t) = -\mu\partial_t\mathcal{I}(t) * \mathcal{L}^{-1}\left\{K_0\left\{[s(s+\alpha)]^{1/2}r/c\right\}\right\} \tag{6.20}$$

for $t > r/c$.

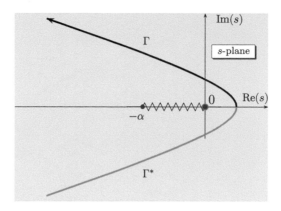

Figure 6.2: Complex s-plane related to the finite-conductivity model. © [2015] IEEE. Adapted, with permission, from [129].

6.2.2 Debye dielectric model

As in the previous subsection 6.2.1, the following considerations are applied to the piecewise linear temporal expansion. Again, referring to Eq. (6.12), the inverse Laplace transformation of the following expressions will be found (cf. Eqs. (6.13) and (6.14))

$$\hat{\gamma}(s)\mathrm{K}_1[\hat{\gamma}(s)r]/s^2 = s[(s+\alpha)/(s+\beta)]^{1/2}$$

$$\mathrm{K}_1\left\{s[(s+\alpha)/(s+\beta)]^{1/2}r/c_\infty\right\}/c_\infty s^2 \qquad (6.21)$$

$$\mathrm{K}_0[\hat{\gamma}(s)r] = \mathrm{K}_0\left\{s[(s+\alpha)/(s+\beta)]^{1/2}r/c_\infty\right\} \qquad (6.22)$$

As no straightforward closed-form inversion of (6.21) and (6.22) exists, the numerical technique as described in Appendix I is applied. In the corresponding complex s-plane we encounter two branch points at $s = \{-\alpha, -\beta\}$ and a double pole at $s = 0$ in Eq. (6.21). Also, the corresponding propagation coefficient $\hat{\gamma}$ shows the inverse-square root singularity at $s = -\beta$ meaning that Eq. (6.21) is unbounded there. The corresponding branch cuts are chosen such that $\mathrm{Re}[(s+\alpha)^{1/2}] \geq 0$ and $\mathrm{Re}[(s+\beta)^{1/2}] \geq 0$ for all $s \in \mathbb{C}$, which implies two overlapping branch cuts along the negative real axis $\{s \in \mathbb{C}; -\infty < \mathrm{Re}(s) \leq -\alpha, \mathrm{Im}(s) = 0\}$, $\{s \in \mathbb{C}; -\infty < \mathrm{Re}(s) \leq -\beta, \mathrm{Im}(s) = 0\}$ together with

$$c_\infty\hat{\gamma}(s) = s + (\alpha - \beta)/2 + \mathcal{O}(s^{-1}) \qquad (6.23)$$

as $|s| \to \infty$. Based on the large-argument expansions (6.18), the Bromwich inversion contour can be closed to the right for $t > r/c_\infty$ and the corresponding integration is, in view of Cauchy's theorem, carried out along the hyperbolic contour $\Gamma \cup \Gamma^*$ (here * denotes complex

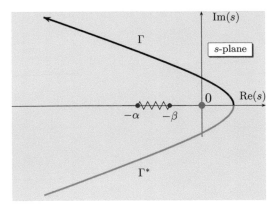

Figure 6.3: Complex s-plane related to the Debye model. © [2015] IEEE. Adapted, with permission, from [129].

conjugate) shown in Fig. 6.3. Owing to the inverse square-root singularity affecting the inversion of Eq. (6.21), the new integration path should not be too close to the singularity at $s = -\beta$. In Fig. 6.3, only the non-overlapping part of the branch cuts is shown. For further details concerning the numerical integration along the hyperbolic contour we refer the reader to Appendix I.

Finally, the unknown expansion coefficients follow upon solving (3.8), provided that the integrated functions ψ and Φ defined in Eqs. (3.13) and (3.14) are replaced by their corresponding (numerical) inversions

$$\psi(r,t) = \mathcal{L}^{-1}\Big\{s[(s+\alpha)/(s+\beta)]^{1/2}$$
$$\mathrm{K}_1\Big\{s[(s+\alpha)/(s+\beta)]^{1/2}r/c_\infty\Big\}/s^2\Big\} \tag{6.24}$$

$$\Phi(r,t) = -\mu\partial_t\mathcal{I}(t) * \mathcal{L}^{-1}\Big\{\mathrm{K}_0\Big\{s[(s+\alpha)/(s+\beta)]^{1/2}r/c_\infty\Big\}\Big\} \tag{6.25}$$

for $t > r/c_\infty$.

6.3 Numerical results

This section is divided into three parts and presents sample numerical calculations concerning the inclusion of radiation loss through the admittance-wall boundary condition and dielectric Boltzmann-type relaxation mechanisms via the finite-conductivity and Debye-type models. All the obtained results are compared with the corresponding results evaluated using FIT of CST Microwave Studio®.

6.3.1 Effect of the admittance wall

In this section, the effect of the admittance boundary is investigated. To this end, we analyse the irregularly-shaped planar circuit as shown in Fig. 3.6b with the instantaneously-reacting dielectric filling and the instantaneously-reacting admittance wall along its rim. The corresponding dielectric relaxation function is impulsive and reads $\kappa(t) = \epsilon\delta(t)$. For the following example we take $\epsilon = 4.5\epsilon_0$ and $d = 1.50\,[\text{mm}]$. Similarly, the TD wall admittance is $\eta_w(\boldsymbol{x},t) = \eta_0\delta(t)$. Here, η_0 is typically a small fraction of the free-space admittance $(\epsilon_0/\mu_0)^{1/2}$ and its specification is a matter of experience. This is, however, a typical feature of including radiation losses in the cavity model. For the presented calculation we take $\eta_0 = 4.0\cdot10^{-3}(\epsilon_0/\mu_0)^{1/2}$ and $\delta = 1.0\cdot10^{-6}ct_w$ (see Fig. 6.1).

The structure is activated via a vertical cylindrical port of radius $1.0\,[\text{mm}]$ with its center placed at $\{x_1^S, x_2^S\} = \{0.03, 0.05\}\,[\text{m}]$. The TD electric-field response is probed at $\{x_1^P, x_2^P\} = \{0.0167, 0.15\}\,[\text{m}]$ within the time window of observation $\{0 \leq ct \leq 3.0\}\,[\text{m}]$. The excitation pulse waveform is shown in Fig. 3.3. The reference FIT model is discretized with a hexahedral mesh and the sidewalls of its surrounding box are defined as 'open' with added space around the structure. The total number of the mesh cells is about 135 thousand.

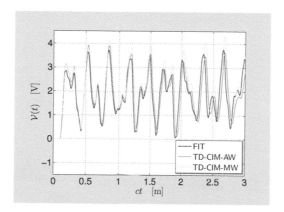

Figure 6.4: The pulsed voltage response evaluated using the proposed TD-CIM and the referential FIT. Radiation loss is accounted for via the admittance-wall (AW) concept and its effect is illustrated on the model with the perfect magnetic wall (MW).

The results are shown in Fig. 6.4. In this figure, we plot the pulsed voltage response evaluated using (a) FIT as implemented in CST Microwave Studio®; (b) TD-CIM with the admittance-wall (AW) condition (viz. Eq. (6.2)); (c) TD-CIM with the classic perfect magnetic-wall (MW) condition (viz. Eq. (2.6)). The improvement brought about by the implementation of the instantaneously reacting admittance-wall condition is obvious. Even better results can be expected for a more sophisticated wall admittance.

6.3.2 Inclusion of conduction loss

The present subsection is related to Sec. 6.2.1 and provides a sample numerical result concerning a rectangular planar circuit with the conductive and dielectric filling and with the perfect magnetic wall along its rim. The EM properties of the slab are now defined via its electric permittivity $\epsilon = 2.50\epsilon_0$ and electric conductivity $\sigma = 0.02\,[\mathrm{S/m}]$. Its thickness is again $d = 1.50\,[\mathrm{mm}]$. The model of the circuit as used for the TD-CIM simulation is shown in Fig. 6.5.

As the excitation pulse we take, again, the waveform shown in Fig. 3.3. The structure is activated via a vertical cylindrical port of radius $1.0\,[\mathrm{mm}]$ with its center placed at $\{x_1^S, x_2^S\} = \{0.0250, 0.0375\}\,[\mathrm{m}]$. The time-domain electric-field response is probed at (a) $\{x_1^P, x_2^P\} = \{0.100, 0.075\}\,[\mathrm{m}]$ (**PROBE 1**) and (b) $\{x_1^P, x_2^P\} = \{0.075, 0.125\}\,[\mathrm{m}]$ (**PROBE 2**) both within the time window of observation $\{0 \leq ct \leq 3.0\}\,[\mathrm{m}]$. The discretization of circuit's rim is uniform with the line segment of length $|\triangle\Omega^{[n]}| = 0.0125\,[\mathrm{m}]$, which corresponds to a eighth of the excitation pulse spatial support $ct_\mathrm{w} = 0.10\,[\mathrm{m}]$. The reference FIT model is composed of about 65 thousand mesh cells and its lateral sides are defined as the perfect magnetic wall. The corresponding attenuated pulsed voltage responses are shown in Figs. 6.6a and 6.6b. The obtained TD-CIM- and FIT-based signals are almost identical.

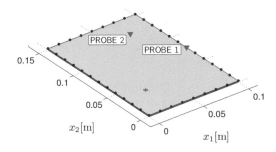

Figure 6.5: Computational model of the analysed circuit with probes (the solid triangles) and a vertical excitation port (the dots on the patch).

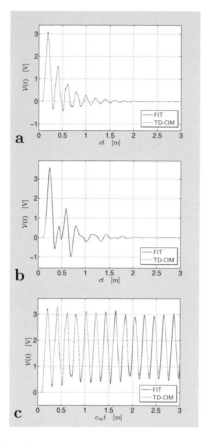

Figure 6.6: The pulsed voltage responses as evaluated using the proposed TD-CIM and the referential FIT of the rectangular circuit with (a) conduction loss observed at **PROBE 1**; (b) conduction loss observed at **PROBE 2**; (c) Debye's relaxation observed at **PROBE 1**. © [2015] IEEE. Adapted, with permission, from [129].

6.3.3 *Inclusion of Debye's dielectric relaxation*

The numerical example that follows is related to Sec. 6.2.2. The following numerical experiment validates the proposed inclusion of the Debye dielectric relaxation model in TD-CIM. In this example, the EM properties of the slab are defined via its relative electric permittivities $\epsilon_r = 4.410$, $\epsilon_\infty = 4.195$ and relaxation time $\tau_r = 1.630 \cdot 10^{-11}$ [s]. The material parameters correspond to 'FR-4 (lossy)' as defined in the CST Material Library. Except for the material filling, the numerical models used for the TD-CIM and FIT simulations remain the same as in the previous section. The voltage pulses evaluated at

$\{x_1^P, x_2^P\} = \{0.100, 0.075\}$ [m] (**PROBE 1**) are shown in Fig. 6.6c. Evidently the signals resulting from TD-CIM and FIT almost overlap each other. It is worth noting that the early parts of the corresponding pulsed responses for conduction losses (viz Fig. 6.6a) and Debye's relaxation behavior (viz Fig. 6.6c) are almost identical. This observation is in agreement with the fact that any dispersion phenomenon shows itself in a transition region after the relevant wavefront has passed [37].

6.4 Conclusions

In order to incorporate the radiation loss and relaxation behavior of a planar circuit into TD-CIM, two of its extensions have been proposed. The first one introduces the admittance-wall concept via a Dirichlet-to-Neumann boundary condition, relating the electric field component to its normal derivative on the circuit periphery. The numerical results concerning the simplest possible instantaneously-reacting edge admittance have indicated the improvement with respect to the standard model incorporating the magnetic-wall boundary condition. The inclusion of relaxation effects has been accomplished by a dedicated Laplace-transform inversion. The latter makes use of the Bromwich-path deformation in the complex-frequency plane and allows for the inclusion of rather general relaxation mechanisms. The corresponding numerical examples concerning the conduction-loss and Debye-type relaxation functions have shown a very good correspondence with respect to (three-dimensional) FIT.

Chapter 7

Inclusion of Linear Lumped Elements

There are a number of practical applications for which the embedding of a lumped circuit element into the planar circuitry is of high importance. Examples from this category can be found in antenna engineering, for instance, where shorting-post, resistor- or capacitor-loaded antennas offer the possibility of efficient antenna size reduction, broad-banding or/and the radiation-pattern/polarization control (e.g. [99]). Another example from the field of EMC is decoupling capacitors that serve for the switching-noise mitigation on PCBs [36]. Accordingly, having its high practical importance in mind, this chapter is aimed at the embedding of basic lumped elements into the framework of TD-CIM.

The present chapter is organized as follows. At first, starting from the contour-integral reciprocity formulation, a general concept of the lumped-element inclusion in TD-CIM is introduced. The subsequent sections address the inclusion of basic linear lumped elements. Namely, the embedding of an inductor, a resistor and a capacitor are each described in detail. It is shown that the effect of an external circuit element shows itself as a separate square 2D-array in the resulting system of equations. This formulation hence makes it possible to evaluate the impact of inclusions separately without having to evaluate the bare structure over again in a new simulation run. Finally, the pulsed voltage responses as observed across the embedded elements are evaluated and validated using FIT.

7.1 General formulation

The point of departure for our analysis is the the first term on the right-hand side of the reciprocity relation (2.34), namely

$$\int_{\boldsymbol{x}\in\Omega} \hat{E}_3^B(\boldsymbol{x}|\boldsymbol{x}^S,s)\hat{J}_3(\boldsymbol{x},s)\mathrm{d}A(\boldsymbol{x}) \tag{7.1}$$

where $\hat{E}_3^B$ is given in Eq. (2.35) and $\hat{J}_3$ is the vertical electric-current density flowing across a lumped element connected between the top and bottom planes (see Fig. 7.1a) via the circular port bounded with $\partial\Omega^C \subset \partial\Omega$ and $\Omega^C = \mathrm{supp}[\hat{J}_3(\boldsymbol{x},s)]$. The circular port is described by its radius ϱ and its center is located at $\boldsymbol{x} = \boldsymbol{x}^C$. As the radius of the port is assumed to be small with respect to the spatial width of the excitation pulse, i.e. $\varrho \ll ct_{\mathrm{w}}$, we can assume the constant electric-current distribution over the port's boundary surface

$$\hat{J}_3(\boldsymbol{x},s) \simeq \partial\hat{J}_3^C(s)\delta(\boldsymbol{x}-\boldsymbol{x}') \quad \text{for } \boldsymbol{x}' \in \partial\Omega^C \tag{7.2}$$

Next, taking into the account the lossless causal fundamental solution [69, Sec. 11.2] together with Eq. (2.35), Eq. (7.1) can be rewritten as

$$\left[-s\mu\partial\hat{J}_3^C(s)/2\pi\right] \int_{\boldsymbol{x}^T\in\partial\Omega} \partial\hat{J}_3^B(\boldsymbol{x}^T|\boldsymbol{x}^S,s)$$
$$\int_{\boldsymbol{x}\in\partial\Omega^C} \mathrm{K}_0[sr(\boldsymbol{x}|\boldsymbol{x}^T)/c]\mathrm{d}l(\boldsymbol{x})\mathrm{d}l(\boldsymbol{x}^T) \tag{7.3}$$

The inner integral in Eq. (7.3) can be evaluated analytically for the circular domain. With the aid of the addition theorems for Bessel functions (see [111, Sec. 6.11] and [1, (9.6.3), (9.6.4)]) we end up with

$$\int_{\boldsymbol{x}\in\partial\Omega^C} \mathrm{K}_0[sr(\boldsymbol{x}|\boldsymbol{x}^T)/c]\mathrm{d}l(\boldsymbol{x}) = 2\pi\varrho\,\mathrm{I}_0(s\varrho/c)\mathrm{K}_0[sr(\boldsymbol{x}^C|\boldsymbol{x}^T)/c] \tag{7.4}$$

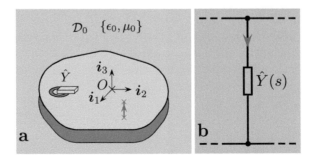

Figure 7.1: Problem configuration. (a) Planar circuit with a linear lumped element; (b) admittance $\hat{Y}(s)$ connected between the conducting planes.

where $I_0(x)$ is the modified Bessel function of the first kind and of the zeroth order. In the next step, the electric-current surface density $\partial \hat{J}_3^C(s)$ is related to the corresponding electric-field strength $\hat{E}_3^C(s)$ via the complex-FD admittance $\hat{Y}(s)$ according to

$$\partial \hat{J}_3^C(s) = \hat{E}_3^C(s)\hat{Y}(s)\,d/2\pi\varrho \tag{7.5}$$

and Eq. (7.4) for $\varrho \downarrow 0$ is used in Eq. (7.3), which leads to

$$-\frac{s\mu d}{2\pi}\hat{E}_3^C(s)\hat{Y}(s)\int_{\boldsymbol{x}^T \in \partial\Omega} \partial \hat{J}_3^B(\boldsymbol{x}^T|\boldsymbol{x}^S, s)\mathrm{K}_0[sr(\boldsymbol{x}^C|\boldsymbol{x}^T)/c]\mathrm{d}l(\boldsymbol{x}^T) \tag{7.6}$$

It is worth noting that for a finite radius of the circular port $\boldsymbol{x}^C \neq \boldsymbol{x}^T$ for all $\boldsymbol{x}^T \in \partial\Omega^C \subset \partial\Omega$, which avoids the necessity of handling the singular self-coupling terms.

In order to solve the problem numerically, the boundary contour $\partial\Omega$ is next approximated by a set of line segments according to Eq. (3.3) and the unknown electric-field strength is expanded through a set of the triangular functions (cf. Eq. (3.4)), that is

$$\hat{E}_3^C(s) = \sum_{k=1}^{NT} e_{[k]}^{[C]}\hat{T}_{[k]}(s) \tag{7.7}$$

The testing electric-current surface density is taken to be a piecewise linear function in space $\partial \hat{J}_3^B(\boldsymbol{x}|\boldsymbol{x}^S, s) = T^{[S]}(\boldsymbol{x})$ for all nodal points $S = \{1, ..., N\}$. Recall that the testing procedure also includes the discretization nodes on the port's periphery $\partial\Omega^C$ that are associated with a finite set $\Pi \subset \{1, ..., N\}$. Making use of the expansion and testing functions, Eq. (7.6) is cast into its matrix form whose TD counterpart is subsequently substituted in the starting reciprocity relation (2.34). In this way, we will end up with a space-time system of equations that has the same form as Eq. (6.4) and can for a loss-free planar circuit, be written as

$$\left(\boldsymbol{I} - \boldsymbol{Q}_{[0]} + \boldsymbol{L}_{[0]}\right) \cdot \boldsymbol{E}_{[p]} = \sum_{k=1}^{p-1}\left(\boldsymbol{Q}_{[p-k]} - \boldsymbol{L}_{[p-k]}\right) \cdot \boldsymbol{E}_{[k]} + \boldsymbol{F}_{[p]} \tag{7.8}$$

Here, apparently, the presence of lumped elements is accounted for via a $[N \times N \times NT]$ 3D-array $\boldsymbol{L}$. A detailed description of its elements concerning the basic lumped elements, such as an inductor, a resistor and a capacitor shall be given in the following section. For sample MATLAB® implementations concerning the evaluation of the lumped-element array $\boldsymbol{L}$ we refer the reader to Appendix E. Finally, the elements of $\boldsymbol{I}$, $\boldsymbol{Q}_{[p-k]}$ and $\boldsymbol{F}_{[p]}$ can be evaluated via Eqs. (3.9), (3.10) and (3.11), respectively, with the TD functions given in Eqs. (3.12)–(3.14).

7.2   Inclusion of a resistor

The resistor is a passive lumped element that can be characterized by its resistance R, which is a real-valued and positive constant and the inverse of conductance G. The corresponding complex–FD admittance is

$$\hat{Y}(s) = G \qquad (7.9)$$

The non-zero array elements describing the presence of a resistor then read (cf. Eq. (6.5))

$$(\boldsymbol{L}_{[p-k]})_{S,m} = \frac{1}{\pi} \frac{G}{\eta} \frac{d}{c\triangle t} \int_{\boldsymbol{x}^T \in \partial\Omega} T^{[S]}(\boldsymbol{x}^T)$$
$$\Theta[r(\boldsymbol{x}^C|\boldsymbol{x}^T), (p-k)\triangle t] \mathrm{d}l(\boldsymbol{x}^T) \qquad (7.10)$$

for all $S = \{1, ..., N\}$, $m \in \Pi$, $t \in \mathcal{T}$, where $\eta = (\epsilon/\mu)^{1/2}$ and

$$\Theta(r, t) = \zeta(r, t + \triangle t) - 2\zeta(r, t) + \zeta(r, t - \triangle t) \qquad (6.6 \text{ revisited})$$
$$\zeta(r, t) = \ln\left[ct/r + (c^2 t^2/r^2 - 1)^{1/2}\right] \mathrm{H}(t - r/c) \qquad (6.7 \text{ revisited})$$

The lumped-element array $\boldsymbol{L}$ corresponding to an inductor and a capacitor will be constructed in a similar way.

7.3   Inclusion of an inductor

The inductor is a passive lumped element that can be characterized by its inductance L, which is a real-valued and positive constant. The corresponding complex-FD admittance then reads

$$\hat{Y}(s) = 1/sL \qquad (7.11)$$

The non-zero array elements describing the presence of an inductor then read

$$(\boldsymbol{L}_{[p-k]})_{S,m} = \frac{1}{\pi} \frac{\mu d}{L} \int_{\boldsymbol{x}^T \in \partial\Omega} T^{[S]}(\boldsymbol{x}^T)$$
$$\Theta[r(\boldsymbol{x}^C|\boldsymbol{x}^T), (p-k)\triangle t] \mathrm{d}l(\boldsymbol{x}^T) \qquad (7.12)$$

for all $S = \{1, ..., N\}$, $m \in \Pi$ and $t \in \mathcal{T}$, where $\Theta(r, t)$ is modified accordingly, viz

$$\Theta(r, t) = \zeta(r, t + \triangle t) - \zeta(r, t) \qquad (6.9 \text{ revisited})$$

7.4 Inclusion of a capacitor

The capacitor is a passive lumped element that can be characterized by its capacitance C, which is a real-valued and positive constant. The corresponding complex–FD admittance is

$$\hat{Y}(s) = sC \tag{7.13}$$

The non-zero array elements describing the presence of a capacitor then read

$$(\boldsymbol{L}_{[p-k]})_{S,m} = \frac{1}{\pi}\frac{C}{\epsilon}\frac{d}{(c\triangle t)^2}\int_{\boldsymbol{x}^T\in\partial\Omega} T^{[S]}(\boldsymbol{x}^T)$$
$$\Theta[r(\boldsymbol{x}^C|\boldsymbol{x}^T),(p-k)\triangle t]\mathrm{d}l(\boldsymbol{x}^T) \tag{7.14}$$

for all $S = \{1, ..., N\}$, $m \in \Pi$, $t \in \mathcal{T}$, where $\Theta(r, t)$ is modified accordingly, viz

$$\Theta(r, t) = \zeta(r, t + \triangle t) - 3\zeta(r, t)$$
$$+ 3\zeta(r, t - \triangle t) - \zeta(r, t - 2\triangle t) \quad \text{(6.11 revisited)}$$

7.5 Numerical results

In this section we will analyse sample planar circuits using TD-CIM and FIT as implemented in CST Microwave Studio®. For the sake of simplicity, we shall analyse loss-free circuits whose dielectric slab has permittivity $\epsilon = 4.50\,\epsilon_0$ and its thickness is $d = 1.20\,[\mathrm{mm}]$. The circuits under consideration are activated by the bell-shaped pulse shown in Fig. 3.3 via a vertical excitation port of radius $1.0\,[\mathrm{mm}]$. Similarly, lumped elements are connected via the circular ports whose radius is $\varrho = 1.0\,[\mathrm{mm}]$. Note that the radius is relatively small with respect to the excitation-pulse spatial support, namely $ct_\mathrm{w} = 100\varrho$. The periphery of the excitation and connecting ports is approximated by a hexagonal bounding contour. The resulting pulse shapes are observed within the finite time window of observation specified via $\{0 \le ct \le 3.0\}\,[\mathrm{m}]$.

7.5.1 Rectangular circuit with a single lumped element

In order to adjust the TD-CIM and FIT-based computational models, we first analyse a simple rectangular planar circuit with the magnetic-wall boundary condition (see Eq. (2.6)) imposed along its rectangular periphery and with a single lumped element. The dimensions of the circuit are $L = 0.10\,[\mathrm{m}]$ and $W = 0.15\,[\mathrm{m}]$ along the x_1- and

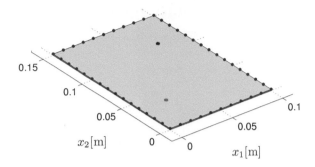

Figure 7.2: Computational model of the analysed rectangular circuit with a vertical excitation port and a lumped element.

x_2-direction, respectively. The vertical excitation port has its center at $\{x_1^S, x_2^S\} = \{L/4, W/4\}$ and the lumped element is placed at $\{x_1^C, x_2^C\} = \{3L/4, 3W/4\}$ (see Fig. 7.2).

For the TD-CIM-based analysis, the circuit periphery is divided into the line segments of length $|\triangle\Omega^{[n]}| = 0.01\,[\mathrm{m}]$, which corresponds to one tenth of the excitation pulse spatial support. The relevant array elements are evaluated using the 6-point Gauss-Legendre quadrature [1, (25.4.30)]. The corresponding FIT model is discretized with a hexahedral mesh. The sidewalls of the surrounding box are defined as the magnetic walls with the vanishing tangential magnetic-field component.

The voltage pulses observed across resistor $G = 0.5\,[\mathrm{S}]$ (i.e. $R = 2.0\,[\Omega]$), inductor $L = 0.10\,[\mathrm{nH}]$ and capacitor $C = 10\,[\mathrm{nF}]$ are shown in Figs. 7.3a–7.3c, respectively. Here, it is observed that for L and C elements the FIT model is very sensitive to the number of mesh lines along the vertical direction. This fact is demonstrated in Figs. 7.3b–7.3c, where one can observe significant deviations in the late-time part of the FIT-based voltage responses for the chosen grid steps $\delta_z = \{d/10, d/50\}$. As the hexahedral FIT-related mesh becomes denser, the TD-CIM-based and FIT-based results converge. The proposed TD-CIM approach, however, has rendered the pulses with significantly less computational efforts. Finally, observe that before the edge-reflected waves arrive to the lumped element at $\boldsymbol{x} = \boldsymbol{x}^C$, the scaled copy, the time derivative and the time integral of the excitation current pulse (see Fig. 3.3) can be recognized in Figs. 7.3a–7.3c, respectively. Afterwards, the total wave field is naturally much more complex.

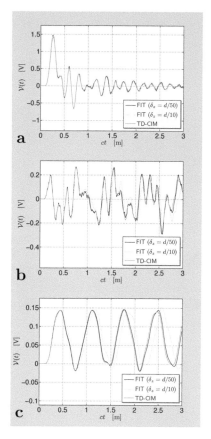

Figure 7.3: The pulsed voltage across the (a) resistor $G = 0.5$ [S]; (b) inductor $L = 0.10$ [nH]; (c) capacitor $C = 10$ [nF], evaluated using the proposed TD-CIM and the referential FIT.

7.5.2 Loaded irregularly-shaped circuit

Having the proper model parameters for the embedding of lumped elements in the numerical models at our disposal, we can now evaluate the TD voltage response of the irregularly-shaped planar circuit given in Fig. 7.4. This structure is activated by the excitation port whose center is at $\{x_1^S, x_2^S\} = \{75.0, 37.5\}$ [mm] and loaded by a resistor $R = 50$ [Ω] at $\{x_1^S, x_2^S\} = \{50.0, 37.5\}$ [mm] and two capacitors of $C = 10$ [nF] placed at $\{x_1^S, x_2^S\} = \{50.0, 93.75\}$ [mm] and $\{x_1^S, x_2^S\} = \{25.0, 93.75\}$ [mm].

Figure 7.5a shows the resulting pulsed voltage response as observed across the 50 [Ω] load resistance. It is seen that the pulse shapes evaluated using TD-CIM and FIT agree well. Again, the suggested methodol-

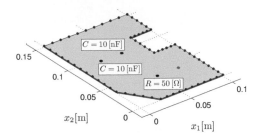

Figure 7.4: Computational model of the analysed irregularly-shaped circuit with a vertical excitation port and three lumped loads.

ogy proves to be computationally more efficient with respect to the referential technique. In order to demonstrate the effect of the capacitors, the pulsed response has also been evaluated without the capacitors. The corresponding pulse is shown in Fig. 7.5b. As can be observed, the capacitors have reduced the signal's average as well as its peak value. On the other hand, the rising edge of the response is apparently not affected by the capacitors.

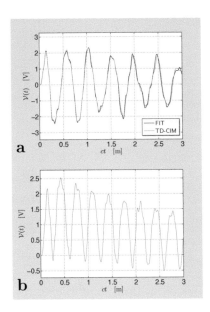

Figure 7.5: The pulsed voltage across the resistor $R = 50\,[\Omega]$ (a) with the two capacitors $C = 10\,[\text{nF}]$ connected; (b) without the capacitors.

7.6   Conclusions

The incorporation of lumped elements in the TD-CIM framework has been described. It has been shown that a linear lumped element can be included in a straightforward way via a separate 2D-array whose elements have been detailed for an inductor, a resistor and a capacitor. As such, this formulation makes it possible to evaluate the effect of external circuit elements separately without having to re-analyze the bare planar structure in a new simulation run. The numerical results concerning the pulsed voltage response evaluated across the elements have shown a good correspondence with respect to (three-dimensional) FIT.

Chapter 8

Far-field Radiation Characteristics

Modeling of pulsed EM radiation is of interest for the TD performance analysis of microstrip antennas as well as for the evaluation of unintentional EMI due to noise signals propagating on PCBs. In the case of a planar structure, one may take advantage of its (relatively) low thickness and make use of a simplified radiation model. In that model, the electric-current surface density on conducting plates is neglected against the magnetic-current surface density along the lateral sides of a planar circuit. In this way, the radiated EM field can be represented via the slant-stack transformation of the equivalent magnetic-current surface density distributed along the circuit's rim. This approximation works well for thin planar circuits and was formerly applied to the FD radiation-field analysis of patch antennas (e.g. [61]) as well as power-ground structures [136]. The present chapter aims to develop the corresponding TD formulation.

The ensuing sections are organized as follows. Firstly, in Sec. 8.1, the electric-field and magnetic-field radiation characteristics are represented via a line integral of the tangential magnetic-current surface density on the circuit periphery. Secondly, in the following Sec. 8.2, upon incorporating the piecewise linear expansion of the equivalent magnetic-current surface density, the radiation integral is approximated by a sum of the nodal electric-field coefficients resulting from TD-CIM. In Sec. 8.3, calculated pulse shapes of the relevant pulsed EM radiation characteristics are compared with the ones evaluated using FIT. Finally, time-varying three-dimensional radiation patterns of an irregularly-shaped planar antenna are presented.

8.1 Radiation model of a planar circuit

We shall analyse the pulsed EM radiation from an irregularly-shaped planar circuit that is placed in the homogeneous and isotropic embedding $\mathcal{D}_0$. The latter is described by its electric permittivity ϵ_0 and magnetic permeability μ_0 (see Fig. 8.1). The corresponding EM wave speed is $c_0 = (\epsilon_0\mu_0)^{-1/2}$ with $(.)^{1/2} > 0$.

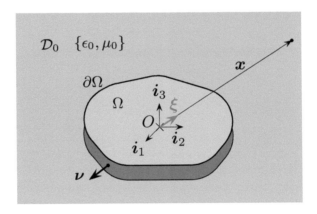

Figure 8.1: Radiation from a planar circuit. © [2014] IEEE. Reprinted, with permission, from [125].

For planar circuits whose thickness d is negligible with respect to the spatial support of an excitation pulse, the total radiated EM field is predominantly generated by the magnetic-current surface density distributed on the lateral sides of the planar circuit [4, Sec. 14.2.2]. For field points placed far from a circuit, the EM field components admit the far-field expansion [24, Sec. 26.12]

$$\{\boldsymbol{E}, \boldsymbol{H}\}(\boldsymbol{x}, t) = \frac{\{\boldsymbol{E}_\infty, \boldsymbol{H}_\infty\}(\boldsymbol{\xi}, t - |\boldsymbol{x}|/c_0)}{4\pi|\boldsymbol{x}|} \left[1 + \mathcal{O}\left(|\boldsymbol{x}|^{-1}\right)\right] \quad (8.1)$$

as $|\boldsymbol{x}| \to \infty$ for $\{t \in \mathbb{R}; t > 0\}$, where $\{\boldsymbol{E}_\infty, \boldsymbol{H}_\infty\}$ are the electric-field and magnetic-field vector radiation characteristics, respectively, and $\boldsymbol{\xi} = \boldsymbol{x}/|\boldsymbol{x}|$ is the unit vector in the direction of observation. Note that, in contrast to the previous sections, all bold Greek and Latin symbols stand for three-dimensional vectors. Interpreting the lateral sides of a circuit as Huygens' surfaces, the electric-field radiation characteristic is represented as

$$\mathbf{I}_t \, \boldsymbol{E}_\infty(\boldsymbol{\xi}, t) = \frac{d}{c_0} \, \boldsymbol{\xi} \times \int_{\boldsymbol{x}' \in \partial\Omega} E_3(\boldsymbol{x}', t + \boldsymbol{\xi} \cdot \boldsymbol{x}'/c_0) \boldsymbol{\tau}(\boldsymbol{x}') \mathrm{d}l(\boldsymbol{x}') \quad (8.2)$$

where the integrand is assumed to be constant along the x_3-direction and $\boldsymbol{\tau} = i_3 \times \boldsymbol{\nu}$ is the unit vector tangential with respect to the circuit periphery $\partial\Omega$. It can be easily verified that the radiation characteristics are interrelated via the plane-wave relation

$$\boldsymbol{\xi} \times \boldsymbol{E}_\infty(\boldsymbol{\xi}, t) = (\mu_0/\epsilon_0)^{1/2} \boldsymbol{H}_\infty(\boldsymbol{\xi}, t) \quad (8.3)$$

Once the space-time distribution of the vertical electric field strength is known, the TD radiation characteristics of a planar circuit can be evaluated according to (8.2) and (8.3) for a given direction, specified via the observation angle $\boldsymbol{\xi}$.

8.2 Evaluation of the radiation integral

Except for special cases, the radiation integral given in Eq. (8.2) has to be evaluated numerically. As far as TD-CIM-based calculations are concerned, the vertical electric-field strength is offered in terms of the (space-time) coefficients along the discretized circuit rim and the time axis. Specifically, the electric-field strength is represented as a piecewise linear function of space (cf. Eqs. (3.4)–(3.5)), that is

$$E_3(\boldsymbol{x}, t) = \sum_{m=1}^{N} e^{[m]}(t) T^{[m]}(\boldsymbol{x}) \quad (8.4)$$

Owing to the fact that the integration of a piecewise linear function along a line segment can be carried out analytically, the spatial integral in (8.2) can be approximately written as the sum of the nodal field coefficients, i.e.

$$\mathbf{I}_t \, \boldsymbol{E}_\infty(\boldsymbol{\xi}, t) \simeq (d/2c_0) \sum_{m=1}^{N} \triangle\Omega^{[m]} (\boldsymbol{\xi} \times \boldsymbol{\tau}^{[m]})$$
$$\left[e^{[m]}(t + \boldsymbol{\xi} \cdot \boldsymbol{x}_c^{[m]}/c_0) + e^{[m+1]}(t + \boldsymbol{\xi} \cdot \boldsymbol{x}_c^{[m]}/c_0) \right] \quad (8.5)$$

where $e^{[m]}(t)$ and $e^{[m+1]}(t)$ are the time-dependent nodal values at the endpoints of the m-th segment, $\boldsymbol{\tau}^{[m]} = (\boldsymbol{x}^{[m+1]} - \boldsymbol{x}^{[m]})/|\boldsymbol{x}^{[m+1]} - \boldsymbol{x}^{[m]}|$ is the unit vector tangential to the m-th segment and finally, $\boldsymbol{x}_c^{[m]} = (\boldsymbol{x}^{[m+1]} + \boldsymbol{x}^{[m]})/2$ localizes its central point. Since the computation runs along a discrete time axis only, the nodal values in (8.5) are appropriately interpolated between the neighboring instants.

8.3 Numerical results

In this section, the pulsed EM radiation characteristics of an irregularly-shaped planar circuit are calculated. For the sake of simplicity, we assume a loss-free dielectric slab of thickness $d = 1.50\,[\text{mm}]$ showing the electric permittivity $\epsilon = 4.0\epsilon_0$ and magnetic permeability $\mu = \mu_0$. The TD-CIM model of the circuit with its discretized rim is shown in Fig. 8.2. The circuit is activated by the electric-current bell-shaped pulse with the amplitude $A = 1.0\,[\text{A}]$ and the pulse time width $ct_\text{w} = 0.10\,[\text{m}]$ (see Fig. 3.3). The pulse is applied to a vertical excitation probe placed at $\{x_1^S, x_2^S\} = \{0.075, 0.05\}\,[\text{m}]$. Consequently, the electric-field radiation characteristics are observed within the time window of observation specified via $\{0 \le c_0 t < 3.0\}\,[\text{m}]$. For validation purposes, we make use of FIT as implemented in the CST Microwave Studio®.

For the TD-CIM model, we assume the vertical excitation port with a hexagonal cross-section of circumradius $0.10\,[\text{mm}]$ and the magnetic wall along the circuit periphery. The total number of the discretization line elements is 58 and the length of all discretization segments is less than one tenth of the spatial width of the excitation pulse. More precisely, we have $\max_m \left(|\Delta\Omega^{[m]}|\right)/ct_\text{w} \simeq 0.083$. On the other hand, the reference FIT model consists of a circular excitation port with radius $0.1\,[\text{mm}]$ and the 'open' boundary condition on the surrounding box. The total number of mesh cells is about 70 thousand.

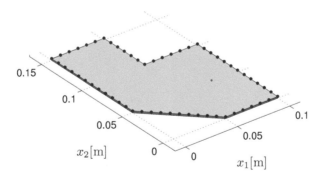

Figure 8.2: Computational model of the analysed circuit with an electric-current vertical port (the dot on the patch).

The electric-field radiation characteristics are evaluated with the help of Eq. (8.5) with the far-field origin (the phase center) placed at $\{0.050, 0.075, 0\}$ [m]. At first, for a given observation angle $\boldsymbol{\xi}$, the contributions from the line segments on the right-hand side of Eq. (8.5) are added up and secondly, the time difference is taken in order to get values of $\boldsymbol{E}_\infty(\boldsymbol{\xi}, t)$. For the sake of convenience, the TD radiation characteristics are expressed in terms of the components with respect to the spherical coordinate system $\{0 \leq R < \infty, 0 \leq \phi \leq 2\pi, 0 \leq \theta \leq \pi\}$

$$E_{\infty;\theta} = E_{\infty;1} \cos(\phi) \cos(\theta) + E_{\infty;2} \sin(\phi) \cos(\theta) - E_{\infty;3} \sin(\theta) \quad (8.6)$$
$$E_{\infty;\phi} = -E_{\infty;1} \sin(\phi) + E_{\infty;2} \cos(\phi) \quad (8.7)$$

where $E_{\infty;k}$ for $k = \{1, 2, 3\}$ denotes a Cartesian component of the three-dimensional radiation vector. Then the direction of observation can be uniquely determined by the spherical angles, i.e. $\boldsymbol{\xi} = \boldsymbol{\xi}(\phi, \theta)$.

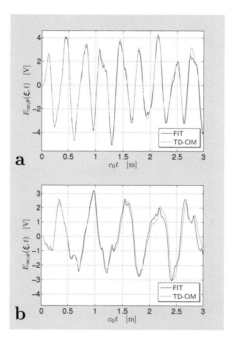

Figure 8.3: The radiated pulse shapes evaluated using the proposed TD-CIM and the referential FIT at $\{\phi, \theta\} = \{\pi/4, \pi/4\}$ for the (a) θ-component; (b) ϕ-component of the vectorial electric-field radiation characteristic.

The pulse shapes of the θ- and ϕ-components of the radiation vector are shown in Figs. 8.3 and 8.4 for two observation angles $\{\phi, \theta\} = \{\pi/4, \pi/4\}$ and $\{\phi, \theta\} = \{0, \pi/3\}$, respectively. As can be observed, the results calculated using TD-CIM and FIT correlate well at the early part of the response and start to deviate at later observation times. Such a discrepancy can be expected due to the different nature of the boundary conditions imposed along the circuit periphery. While the TD-CIM model assumes the perfect magnetic wall here, the circuit boundary in the (three-dimensional) FIT model is fully 'open'. Note that the same behavior was observed in Sec. 3.4, where the vertical component of the electric-field strength along $\partial\Omega$ is evaluated. The latter, in fact, corresponds to the equivalent magnetic-current surface density that generates the observed far-field amplitudes.

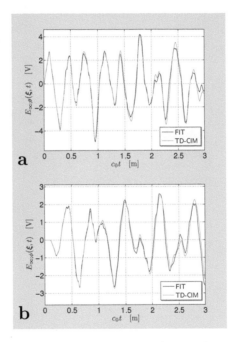

Figure 8.4: The radiated pulse shapes evaluated using the proposed TD-CIM and the referential FIT at $\{\phi, \theta\} = \{0, \pi/3\}$ for the (a) θ-component; (b) ϕ-component of the vectorial electric-field radiation characteristic.

In the next step, three-dimensional, time-varying radiation patterns were calculated. Their surface was constructed, for a fixed point in time, by mapping the absolute values of the far field to a set of observation points on the unit sphere. Consequently, the actual values of the radiated field were represented with a color scale. The results are shown in Figs. 8.5–8.7 for the θ-, ϕ-components and absolute values of the electric-field radiation characteristic, respectively. In order to demon-

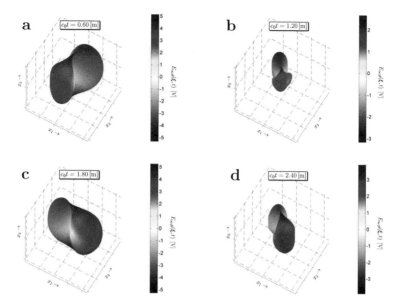

Figure 8.5: Time-varying radiation diagram of the θ-component of the vectorial electric-field radiation characteristic.

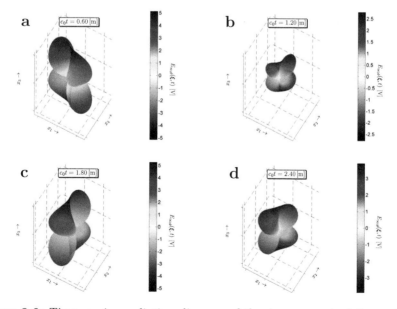

Figure 8.6: Time-varying radiation diagram of the ϕ-component of the vectorial electric-field radiation characteristic.

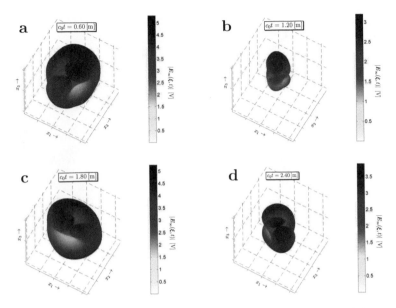

Figure 8.7: Time-varying radiation diagram of the absolute value of the vectorial electric-field radiation characteristic.

strate its time evolution, we take $c_0 t = \{0.60, 1.20, 1.80, 2.40\}$ [m] as the observation time points.

8.4   Conclusions

The pulsed EM radiation from a planar circuit has been analysed in the context of TD-CIM. It has been shown that the electric-field (time-integrated) far-field amplitude can be expressed using the slant-stack transformation of the equivalent magnetic-current surface density distributed along circuit's rim. A straightforward numerical solution of the integral representation has been proposed and numerically validated for the case of an irregularly-shaped planar circuit. The resulting radiated pulse shapes have shown a very good correspondence with respect to the ones evaluated with the aid of (three-dimensional) FIT.

Chapter 9

Time-domain Mutual Coupling between Planar Circuits

Mutual EM coupling effects between microstrip elements have been intensively studied in the real-FD (see e.g. [49, 63, 75, 87]). In this category, two main approaches can be distinguished. The first one is based on the concept of reaction [100] combined with the cavity model [61] and its description in terms of the eigenfunction field expansion [63, 87]. The second approach is somewhat more general and relies on a full wave solution via the method of moments [75, 89].

Despite the fact that high-speed digital circuits inherently operate in TD, the corresponding TD analysis, that would shed light on the mutual interaction of parallel-plate radiators, is very rare in the literature on the subject. Accordingly, the main purpose of this chapter is to fill this void and provide and an efficient methodology for the EM interference analysis of planar circuits directly in space-time. To this end, the reciprocity theorem of the time-convolution type (see [24, Sec. 28.2] and [133, Sec. 1.4.1]) is applied, which yields an efficient computational scheme that serves the purpose very well.

An interesting application of the introduced methodology is the design of intra- and inter-chip wireless interconnections [14, 54] for which the TD EM coupling is primarily a desired phenomenon. In fact, a complete mastery of the pulsed-field transfer between two EM radiators is the key point towards the full utilization of wireless interconnects in practice. Here, too, compliance with the international regulations on EMI is of crucial importance [55]. For a relevant TD reciprocity-based analysis of a two antenna configuration we refer the reader to [107].

The following sections are organized as follows. For the sake of completeness, the starting reciprocity-based contour-integral formulation for the TD-CIM analysis of a single planar circuit is given in Sec. 9.2. Subsequently, the main results of this chapter are introduced in Sec. 9.3 via the systematic application of the reciprocity theorem of the time-convolution type. Here it is shown that the pulsed-voltage response of a receiving (or victim) planar circuit can be expressed via straightforward (rim-to-rim) integral relations consisting of two contour integrations over transmitting- and receiving-circuits' contours. Finally, the resulting coupling model is implemented and validated in Sec. 9.4, where EM coupling of two irregularly-shaped planar circuits is numerically analysed and validated via FIT.

9.1 EM coupling model

The problem configuration consists of a transmitting planar circuit (the emitter) $\mathcal{D}^T \subset \mathbb{R}^3$ and a receiving planar circuit (the susceptor) $\mathcal{D}^R \subset \mathbb{R}^3$ that are placed in the linear, isotropic, homogeneous and loss-free embedding $\mathcal{D}_0$ whose EM properties are described by its electric permittivity ϵ_0 and magnetic permeability μ_0 (see Fig. 9.1). The corresponding EM wave speed is $c_0 = (\epsilon_0 \mu_0)^{-1/2} > 0$. Each planar circuit is described by its own surface domain $\Omega^{T,R} \subset \mathbb{R}^2$, its thickness $d^{T,R}$ and by the relative electric permittivity $\epsilon_r^{T,R}$ of the dielectric filling. The corresponding EM wave speed in the dielectric layer is then $c^{T,R} = c_0(\epsilon_r^{T,R})^{-1/2}$. The outer unit vector with respect to the circuit rims $\partial\Omega^{T,R} \subset \mathbb{R}$ is denoted by $\boldsymbol{\nu}^{T,R} = \boldsymbol{\nu}^{T,R}(\boldsymbol{x})$ and $\boldsymbol{\tau}^{T,R} = \boldsymbol{\tau}^{T,R}(\boldsymbol{x}) = \boldsymbol{i}_3 \times \boldsymbol{\nu}^{T,R}$.

The main goal of the following EM interference analysis is to find the pulsed-voltage response $\mathcal{V}^R$ induced in the receiving circuit $\mathcal{D}^R$ due to the pulsed electric-current excitation $\mathcal{I}^T$ applied to the transmitting circuit $\mathcal{D}^T$.

9.2 A single planar circuit

In accordance with Sec. 2.1.2, a global reciprocity relation is formulated for a single planar circuit. Here we do not distinguish between the transmitting and the receiving circuit. As a consequence, Ω and $\partial\Omega$

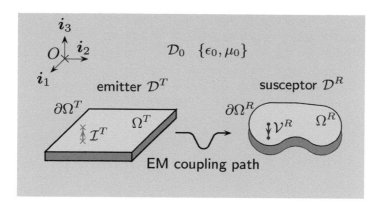

Figure 9.1: Problem configuration consisting of two planar circuits that are electromagnetically coupled. © [2014] IEEE. Reprinted, with permission, from [125].

stand either for $\Omega^{T,R}$ and $\partial\Omega^{T,R}$, respectively. The reciprocity theorem of the time-convolution type is applied to the actual and testing (B) wave fields and to the surface domain Ω of a planar circuit. Applying the testing source density along the circuit periphery only, we finally end up with (cf. Eq. (2.34))

$$\frac{1}{2}\int_{\boldsymbol{x}\in\partial\Omega} E_3(\boldsymbol{x},t) * \partial J_3^B(\boldsymbol{x}|\boldsymbol{x}^S,t)\mathrm{d}l(\boldsymbol{x})$$

$$= \int_{\boldsymbol{x}\in\Omega} E_3^B(\boldsymbol{x}|\boldsymbol{x}^S,t) * J_3(\boldsymbol{x},t)\mathrm{d}A(\boldsymbol{x})$$

$$+ \int_{\boldsymbol{x}\in\partial\Omega} E_3(\boldsymbol{x},t) * \boldsymbol{\nu}(\boldsymbol{x}) \cdot \partial\boldsymbol{J}^B(\boldsymbol{x}|\boldsymbol{x}^S,t)\mathrm{d}l(\boldsymbol{x}) \quad (9.1)$$

where the testing fields are linearly related to their source according to (cf. Eqs. (2.35) and (2.36))

$$E_3^B(\boldsymbol{x}|\boldsymbol{x}^S,t) =$$

$$= -\mu_0\partial_t \int_{\boldsymbol{x}^T\in\partial\Omega} G_\infty[r(\boldsymbol{x}|\boldsymbol{x}^T),t] * \partial J_3^B(\boldsymbol{x}^T|\boldsymbol{x}^S,t)\mathrm{d}l(\boldsymbol{x}^T) \quad (9.2)$$

$$\partial J_\kappa^B(\boldsymbol{x}|\boldsymbol{x}^S,t) =$$

$$= -\int_{\boldsymbol{x}^T\in\partial\Omega} \partial_\kappa G_\infty[r(\boldsymbol{x}|\boldsymbol{x}^T),t] * \partial J_3^B(\boldsymbol{x}^T|\boldsymbol{x}^S,t)\mathrm{d}l(\boldsymbol{x}^T) \quad (9.3)$$

for all $\boldsymbol{x}^S \in \partial\Omega$ and $t > 0$. Here, $r(\boldsymbol{x}|\boldsymbol{x}^T) = |\boldsymbol{x} - \boldsymbol{x}^T|$ is the Eucledian distance between positions given by the position vectors $\boldsymbol{x}$ and $\boldsymbol{x}^T$ and $G_\infty(r,t)$ denotes a fundamental solution of the two-dimensional wave equation that satisfies the zero initial conditions and the causality condition. In the interference analysis that follows, Eq. (9.1) is solved numerically for the both transmitting and receiving planar circuits using TD-CIM. The corresponding solution procedure for a loss-free planar circuit is closely described in Chapter 3 and its extension to dispersive planar circuits can be found in Chapter 6.

9.3 Coupling between two planar circuits

In this section, the TD EM mutual coupling between two planar circuits is described with the help of the reciprocity theorem. To this end, the total field in the receiving state (R) is written as the linear superposition of incident (i) and scattered (s) wave fields, i.e.

$$\{\boldsymbol{E}^R, \boldsymbol{H}^R\} = \{\boldsymbol{E}^i + \boldsymbol{E}^s, \boldsymbol{H}^i + \boldsymbol{H}^s\} \tag{9.4}$$

As the testing state (B) we take the testing wave field $\{\boldsymbol{E}^B, \boldsymbol{H}^B\}$ generated by a vertical excitation port

$$\boldsymbol{J}^B(\boldsymbol{x}, t) = \mathcal{I}^B(t)\delta(\boldsymbol{x} - \boldsymbol{x}^S)\boldsymbol{i}_3 \tag{9.5}$$

for $\boldsymbol{x}^S \in \mathcal{D}^R$, where $\mathcal{I}^B$ is the source signature for which $\mathcal{I}^B(t) = 0$ for $t < 0$. The testing wave field satisfies the magnetic-wall boundary condition $\boldsymbol{\tau}^R \cdot \boldsymbol{H}^B(\boldsymbol{x}, t) = 0$ for all $\boldsymbol{x} \in \partial\Omega^R$ and $t > 0$ (cf. Eq. (2.6)). At this point it is noted that in the analysis we neglect multiple scattering between the planar circuits meaning that the testing wave field does not account for the presence of the transmitting circuit. This assumption, however, does not introduce any error up to the instant when a scattered field gets back to the receiver due to the presence of the transmitter. Consequently, the early part of the pulsed voltage response is exact, whatever EM coupling strength. Beyond this limit, in applications where multiple EM scattering effects are decisive, one has to resort to a full-wave EM numerical solver.

Firstly, the reciprocity theorem is applied to the domain exterior, to the receiving planar structure and to the scattered and testing wave fields, which results in

$$\int_{\boldsymbol{x} \in \partial\mathcal{D}^R} \left[\boldsymbol{E}^B(\boldsymbol{x}, t) \stackrel{*}{\times} \boldsymbol{H}^s(\boldsymbol{x}, t) \right.$$
$$\left. - \boldsymbol{E}^s(\boldsymbol{x}, t) \stackrel{*}{\times} \boldsymbol{H}^B(\boldsymbol{x}, t) \right] \cdot \boldsymbol{\nu}^R(\boldsymbol{x}) \mathrm{d}A(\boldsymbol{x}) = 0 \tag{9.6}$$

for all $t > 0$. Substitution of Eq. (9.4) in (9.6) yields

$$\int_{\boldsymbol{x} \in \partial \mathcal{D}^R} \left[\boldsymbol{E}^B(\boldsymbol{x}, t) \stackrel{*}{\times} \boldsymbol{H}^R(\boldsymbol{x}, t) \right. $$

$$\left. - \boldsymbol{E}^R(\boldsymbol{x}, t) \stackrel{*}{\times} \boldsymbol{H}^B(\boldsymbol{x}, t) \right] \cdot \boldsymbol{\nu}^R(\boldsymbol{x}) \mathrm{d}A(\boldsymbol{x})$$

$$= \int_{\boldsymbol{x} \in \partial \mathcal{D}^R} \left[\boldsymbol{E}^B(\boldsymbol{x}, t) \stackrel{*}{\times} \boldsymbol{H}^i(\boldsymbol{x}, t) \right.$$

$$\left. - \boldsymbol{E}^i(\boldsymbol{x}, t) \stackrel{*}{\times} \boldsymbol{H}^B(\boldsymbol{x}, t) \right] \cdot \boldsymbol{\nu}^R(\boldsymbol{x}) \mathrm{d}A(\boldsymbol{x}) \qquad (9.7)$$

for all $t > 0$. Secondly, the reciprocity theorem is applied to the domain occupied by the receiving circuit $\mathcal{D}^R$ and to the total wave fields in both states

$$\int_{\boldsymbol{x} \in \partial \mathcal{D}^R} \left[\boldsymbol{E}^B(\boldsymbol{x}, t) \stackrel{*}{\times} \boldsymbol{H}^R(\boldsymbol{x}, t) - \boldsymbol{E}^R(\boldsymbol{x}, t) \stackrel{*}{\times} \boldsymbol{H}^B(\boldsymbol{x}, t) \right] \cdot \boldsymbol{\nu}^R(\boldsymbol{x}) \mathrm{d}A(\boldsymbol{x})$$

$$= \int_{\boldsymbol{x} \in \mathcal{D}^R} \boldsymbol{J}^B(\boldsymbol{x}, t) \stackrel{*}{\cdot} \boldsymbol{E}^R(\boldsymbol{x}, t) \mathrm{d}V(\boldsymbol{x}) \qquad (9.8)$$

Owing to the thin-slab approximation and the magnetic-wall boundary condition (see Eq. (2.6)) being satisfied by the testing wave field, we arrive, upon combining Eqs. (9.7)–(9.8), at the desired expression

$$\mathcal{V}^R(\boldsymbol{x}^S, t) * \mathcal{I}^B(t) \simeq$$

$$\simeq - \int_{\boldsymbol{x} \in \partial \Omega^R} \mathcal{V}^B(\boldsymbol{x}|\boldsymbol{x}^S, t) * \boldsymbol{\tau}^R(\boldsymbol{x}) \cdot \boldsymbol{H}^i(\boldsymbol{x}, t) \mathrm{d}l(\boldsymbol{x}) \qquad (9.9)$$

for all $\boldsymbol{x}^S \in \Omega^R$ and $t > 0$. In Eq. (9.9), $\mathcal{V}^R$ is the TD voltage induced in the receiving circuit due to the tangential magnetic field strength radiated from the emitting circuit and $\mathcal{V}^B(\boldsymbol{x}|\boldsymbol{x}^S, t)$ represents the TD testing voltage along the rim of the receiving circuit $\boldsymbol{x} \in \partial \Omega^R$ activated via the electric-current point source placed at $\boldsymbol{x}^S \in \Omega^R$ (see Eq. (9.5)). The incident magnetic field strength is given via the following Kirchhoff-Huygens electromagnetic field representations

$$\boldsymbol{H}^i = \boldsymbol{H}^{i;\mathrm{NF}} + \boldsymbol{H}^{i;\mathrm{IF}} + \boldsymbol{H}^{i;\mathrm{FF}} \qquad (9.10)$$

where

$$\boldsymbol{H}^{i;\mathrm{NF}}(\boldsymbol{x}^R, t) \simeq -\mu_0^{-1} \int_{\boldsymbol{x} \in \partial \Omega^T} \frac{\mathrm{I}_t \mathcal{V}^T(\boldsymbol{x}, t - |\boldsymbol{x}^R - \boldsymbol{x}|/c_0)}{4\pi |\boldsymbol{x}^R - \boldsymbol{x}|^3}$$

$$\left[3 \boldsymbol{\xi}(\boldsymbol{x}^R - \boldsymbol{x}) \boldsymbol{\xi}^\mathrm{T}(\boldsymbol{x}^R - \boldsymbol{x}) - \mathbb{I} \right] \cdot \boldsymbol{\tau}^T(\boldsymbol{x}) \mathrm{d}l(\boldsymbol{x}) \qquad (9.11)$$

$$\boldsymbol{H}^{i;\text{IF}}(\boldsymbol{x}^R, t) \simeq -\eta_0 \int_{\boldsymbol{x} \in \partial \Omega^{\text{T}}} \frac{\mathcal{V}^T(\boldsymbol{x}, t - |\boldsymbol{x}^R - \boldsymbol{x}|/c_0)}{4\pi |\boldsymbol{x}^R - \boldsymbol{x}|^2}$$
$$\left[3\boldsymbol{\xi}(\boldsymbol{x}^R - \boldsymbol{x})\boldsymbol{\xi}^{\text{T}}(\boldsymbol{x}^R - \boldsymbol{x}) - \mathbb{I} \right] \cdot \boldsymbol{\tau}^T(\boldsymbol{x}) \mathrm{d}l(\boldsymbol{x}) \qquad (9.12)$$

$$\boldsymbol{H}^{i;\text{FF}}(\boldsymbol{x}^R, t) \simeq -\epsilon_0 \int_{\boldsymbol{x} \in \partial \Omega^{\text{T}}} \frac{\partial_t \mathcal{V}^T(\boldsymbol{x}, t - |\boldsymbol{x}^R - \boldsymbol{x}|/c_0)}{4\pi |\boldsymbol{x}^R - \boldsymbol{x}|}$$
$$\left[\boldsymbol{\xi}(\boldsymbol{x}^R - \boldsymbol{x})\boldsymbol{\xi}^{\text{T}}(\boldsymbol{x}^R - \boldsymbol{x}) - \mathbb{I} \right] \cdot \boldsymbol{\tau}^T(\boldsymbol{x}) \mathrm{d}l(\boldsymbol{x}) \qquad (9.13)$$

for all $\boldsymbol{x}^R \in \partial \Omega^R$ and $t > 0$, where $\mathcal{V}^T$ denotes the voltage distribution on the rim of the transmitting structure, $\boldsymbol{\xi}(\boldsymbol{x}) = \boldsymbol{x}/|\boldsymbol{x}|$ is the unit vector in the direction of $\boldsymbol{x}$, $\mathbb{I}$ is the 3×3 identity matrix, ^{T} denotes the matrix transposition and $\eta_0 = (\epsilon_0/\mu_0)^{1/2} > 0$. Combination of Eqs. (9.9) with (9.10)–(9.13) yields a compact space-time description of the EM mutual coupling between the two interacting planar structures. From these relations we can immediately extract directional terms of the interfering structures that may be used for optimizing their mutual EM coupling strength. For the intermediate/near coupling field this term reads

$$\boldsymbol{\tau}^R(\boldsymbol{x}^R) \cdot \left[3\, \boldsymbol{\xi}(\boldsymbol{x}^R - \boldsymbol{x}^T)\boldsymbol{\xi}^{\text{T}}(\boldsymbol{x}^R - \boldsymbol{x}^T) - \mathbb{I} \right] \cdot \boldsymbol{\tau}^T(\boldsymbol{x}^T) \qquad (9.14)$$

for all $\boldsymbol{x}^{T,R} \in \partial \Omega^{T,R}$. The latter can be, with the aid of the angles depicted in Fig. 9.2, rewritten as

$$3 \cos(\psi^R) \cos(\psi^T) - \cos(\chi^{R;T}) \qquad (9.15)$$

which may, for given circuit's contours $\partial \Omega^{T,R}$ and their mutual orientation, offer the first clue about how to suppress (or eventually stimulate) their EM mutual (intermediate/near) EM-field coupling. A similar expression describing the far-field directional term can be obtained from the combination of Eqs. (9.9) and (9.13).

In summary, the described approach consists of the following steps. Starting with the transmitting structure, the pulsed voltage distribution $\mathcal{V}^T$ is evaluated for all $\boldsymbol{x} \in \partial \Omega^T$ in a given time window and for a given excitation electric-current pulse $\mathcal{I}^T(t)$. This step can be accomplished, for an arbitrarily-shaped planar structure, with the aid of TD-CIM. The evaluated space-time voltage distribution $\mathcal{V}^T$ is subsequently substituted in Eqs. (9.11)–(9.13), which yields the incident magnetic field strength at points along the rim of the receiving structure. Since the domains occupied by the interacting planar circuits

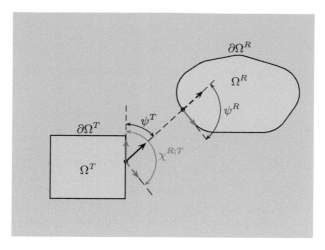

Figure 9.2: Mutually interfering planar circuits (top view) and configurational parameters.

are disjointed Eqs. (9.11)–(9.13) are free of the spatial singularities, and their evaluation, therefore, does not present any difficulties. In the next step, the testing pulsed voltage $\mathcal{V}^B$, due to the electric-current pulse shape $\mathcal{I}^B$, is evaluated for all $\boldsymbol{x} \in \partial\Omega^R$ in a given time window. This step can be carried out with the aid TD-CIM again. Finally, once the time convolution on the right-hand side of Eq. (9.9) is evaluated, $\mathcal{V}^R$ may be readily recovered for a convenient choice of the testing electric-current pulse shape $\mathcal{I}^B(t)$. The easiest way to obtain $\mathcal{V}^R$ from the known convolution $\mathcal{V}^R * \mathcal{I}^B$ would be to choose the Dirac delta test-source signature $\mathcal{I}^B(t) = \delta(t)$. This choice, however, is not possible as the corresponding voltage response $\mathcal{V}^B$ is evaluated using TD-CIM that requires a smoother excitation time signature (see Eq. (3.14)). For the bell-shaped electric-current excitation signature (cf. Eq. (F.6)), i.e.

$$\mathcal{I}^B(t) = 2\left(\frac{t}{t_{\mathrm{w}}}\right)^2 \mathrm{H}(t) - 4\left(\frac{t}{t_{\mathrm{w}}} - \frac{1}{2}\right)^2 \mathrm{H}\left(\frac{t}{t_{\mathrm{w}}} - \frac{1}{2}\right)$$
$$+ 4\left(\frac{t}{t_{\mathrm{w}}} - \frac{3}{2}\right)^2 \mathrm{H}\left(\frac{t}{t_{\mathrm{w}}} - \frac{3}{2}\right) - 2\left(\frac{t}{t_{\mathrm{w}}} - 2\right)^2 \mathrm{H}\left(\frac{t}{t_{\mathrm{w}}} - 2\right) \quad (9.16)$$

that is used in the following numerical examples, one may employ, for example, the following two-step deconvolution procedure. In this approach, the bell-shaped pulse is written out as the time convolution of the rectangular and triangular functions, which readily yields

$$\mathcal{V}^R(\boldsymbol{x}^S, t) = \partial_t^2 \mathcal{T}^T(\boldsymbol{x}^S, t) + 2\mathcal{V}^R(\boldsymbol{x}^S, t - t_{\mathrm{w}}/2) - \mathcal{V}^R(\boldsymbol{x}^S, t - t_{\mathrm{w}}) \quad (9.17)$$

where

$$\mathcal{T}^T(\boldsymbol{x}^S, t) = t_{\mathrm{w}}^2 \partial_t \mathcal{V}^R(\boldsymbol{x}^S, t) * \mathcal{I}^B(t)/4 + \mathcal{T}^T(\boldsymbol{x}^S, t - t_{\mathrm{w}}) \qquad (9.18)$$

corresponds to the time convolution of the sought voltage response $\mathcal{V}^R$ with the triangular pulse $\mathrm{T}(t, t_{\mathrm{w}})$ (see Eq. (F.4)). For an alternative closed-form deconvolution algorithm we refer the reader to Eq. (11.25).

9.4 An illustrative numerical example

This section provides an illustrative numerical example concerning the TD EM mutual coupling between two planar circuits both lying in one plane $x_3 = 0$. The analysed problem configuration is shown in Fig. 9.3. In this configuration, the emitting planar circuit is excited by a vertical electrical-current source localized around $\boldsymbol{x}^T = \{25.0, 37.5, 0\}$[mm] (see the cross symbol in Fig. 9.3). The time signature of the excitation electric-current source is shown in Fig. 3.3. For the sake of simplicity, both analysed planar circuits show the same relative electric permittivity $\epsilon_{\mathrm{r}}^{T,R} = 4.0$, which implies $c = c^R = c^T$. Also, their thickness is $d^{T,R} = 1.50$ [mm].

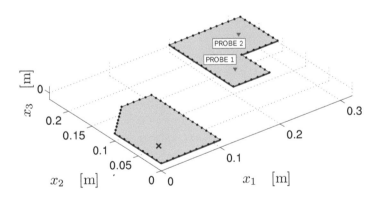

Figure 9.3: Computational model of the analysed configuration in which the transmitting structure is activated by an electric-current vertical port (the cross symbol) and the induced voltage responses are observed by two probes (the solid triangles).

Consequently, the induced pulsed voltage responses are observed at $\boldsymbol{x}^R = \{230, 135, 0\}$ [mm] (**PROBE 1**) and at $\boldsymbol{x}^R = \{290, 205, 0\}$ [mm] (**PROBE 2**) within the receiving planar circuit (see the triangles in Fig. 9.3). To validate the developed computational EM coupling model,

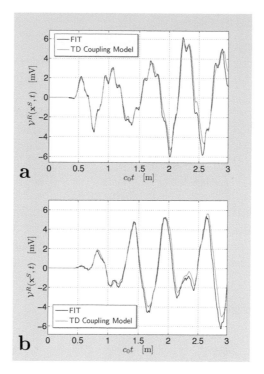

Figure 9.4: The pulsed voltage responses as evaluated using the proposed coupling model and the referential FIT observed at (a) PROBE 1; (b) PROBE 2. © [2014] IEEE. Adapted, with permission, from [128].

the problem has also been solved with the aid of FIT as implemented in CST Microwave Studio®. The corresponding (three-dimensional) FIT-based model consists of about 620 thousand hexahedral mesh cells, while the developed rim-to-rim model requires us to account for only $(48 + 52)$ of the line segments along the peripheral rims (see the black dots in Fig. 9.3). The results are shown in Fig. 9.4. Considering the huge reduction of the solution space, the observed voltage signals correlate very well.

The evaluation of the pulsed-voltage response via the rim-to-rim expressions (9.9)–(9.13) involves only the one-dimensional integrations whatever the emitter/susceptor distance. This property, along with the omission of multiple EM scattering effects, makes the constructed coupling model suitable for applications where the emitter/susceptor distance related to the spatial support of an excitation pulse is relatively large and the traditional direct-discretization numerical approaches require an exceedingly high number of discretization cells within their (truncated) three-dimensional computational domain.

9.5 Conclusions

Space-time mutual EM coupling between two planar circuits has been described in closed form directly in TD. Besides providing physical insights into the mutual (space-time) EM coupling mechanism, the proposed EM coupling model indicates huge savings of computational resources with respect to the traditional direct-discretization computational methods, such as FIT. Consequently, the application of the introduced coupling model is especially profitable in problem configurations where the spatial support of an excitation pulse is relatively short with respect to the emitter/susceptor distance. Thanks to its high computational efficiency, the introduced methodology may find its applications in optimizing TD coupling effects between planar circuits, wireless interconnections and simple antenna arrays. In this respect, an application of the resulting rim-to-rim relations to optimizing the pulsed EM-field transfer between two planar circuits has been hinted at.

Chapter 10

Time-domain Self-reciprocity of a One-port Planar Circuit

The universal property of reciprocity is without a shadow of a doubt among the most intriguing concepts in EM theory. Indeed, the reciprocity theorem furnishes a solid foundation for the general uniqueness theorem [26], encompasses the 'weak formulations' of both direct and inverse scattering/source problems [22] with applications in computational electromagnetics [23] and reveals the fundamental transmission/reception properties of general antenna systems [28, 133].

The vast majority of works on the subject are carried out in the real-FD (see e.g. [21, 27, 93, 113], [122, Sec. 8.7],[15, Sec. 5.1]). Notwithstanding the ever-increasing interest in applications of pulsed EM phenomena such as the use of inter/intra-chip wireless interconnections [14, 55], for example, the TD reciprocity and its EMC/EMI implications are almost untouched upon in the literature on the subject. In this respect, only a few exceptions exist. The general antenna-system description in TD can be found in [5, 28, 108] and a small-antenna UWB radio link and its optimization is investigated in [91, 92], for instance. Accordingly, the main purpose of this chapter is to construct a purely TD self-reciprocity relation that makes it possible to find the pulsed EM radiation characteristics of a planar circuit through its reaction on the incident plane wave in the receiving situation. The results presented in this chapter have been previously discussed in an earlier paper [125] that heavily relies on the seminal paper by De Hoop, Lager and Tomassetti [28].

The ensuing sections are organized as follows. At first, Secs. 10.1–10.3 introduce the reader to the transmitting and receiving situations of a one-port planar circuit. Subsequently, the main results of the chapter are given in Sec. 10.4, where the desired self-reciprocity relation is arrived at via the reciprocity theorem of the time-convolution type [24, Sec. 28.2]. Apart from its application in constructing the pulsed EM radiation characteristics, it is demonstrated in Sec. 10.5 that the introduced self-reciprocity relation is useful for consistency benchmarks of purely numerical EM-field solvers.

10.1   Model definition

Let us analyse a one-port planar circuit that occupies a domain $\mathcal{D} \subset \mathbb{R}^3$ whose bounding surface is denoted by $\partial \mathcal{D} \subset \mathbb{R}^2$. Again, the planar circuit is placed in the linear, isotropic, homogeneous and loss-free embedding $\mathcal{D}_0$ whose EM properties are described by its electric permittivity ϵ_0 and magnetic permeability μ_0. The corresponding EM wave speed is $c_0 = (\epsilon_0 \mu_0)^{-1/2} > 0$. The EM properties of the dielectric slab are described by its electric permittivity ϵ and magnetic permeability μ_0. The electrically conducting plate of the circuit occupies a surface domain $\Omega \subset \partial \mathcal{D}$ bounded by its rim $\partial \Omega \subset \mathbb{R}$. The normal outer unit vector is denoted by $\boldsymbol{\nu}$. Partial differentiation with respect to the spatial coordinates is, for the sake of conciseness, denoted by $\boldsymbol{\nabla} = \partial_1 \boldsymbol{i}_1 + \partial_2 \boldsymbol{i}_2 + \partial_3 \boldsymbol{i}_3$. In the following sections we find a reciprocity relation between transmitting and receiving properties of a one-port planar circuit.

10.2   Transmitting state of a planar circuit

The planar circuit is, in its transmitting situation, excited via the vertical excitation probe (see Fig. 10.1) with defined electric-current density $\boldsymbol{J}^T = J^T \boldsymbol{i}_3$. The emitted ($T$) EM field then satisfies

$$\boldsymbol{\nabla} \times \boldsymbol{H}^T - \epsilon \partial_t \boldsymbol{E}^T = \boldsymbol{J}^T \qquad (10.1)$$

$$\boldsymbol{\nabla} \times \boldsymbol{E}^T + \mu_0 \partial_t \boldsymbol{H}^T = \boldsymbol{0} \qquad (10.2)$$

for all $\boldsymbol{x} \in \mathcal{D}$ and $t > 0$, where the thin-slab approximation applies (cf. Eqs. (2.1)–(2.2)). In the (unbounded) domain exterior to the planar

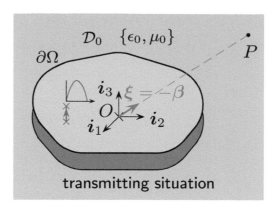

transmitting situation

Figure 10.1: Planar circuit in its transmitting state. © [2014] IEEE. Reprinted, with permission, from [125].

circuit, where the emitted EM field is observed, the governing EM field equations has the following form

$$\boldsymbol{\nabla} \times \boldsymbol{H}^T - \epsilon_0 \partial_t \boldsymbol{E}^T = \boldsymbol{0} \tag{10.3}$$

$$\boldsymbol{\nabla} \times \boldsymbol{E}^T + \mu_0 \partial_t \boldsymbol{H}^T = \boldsymbol{0} \tag{10.4}$$

for all $\boldsymbol{x} \in \mathcal{D}_0$ and $t > 0$ together with the 'radiation condition'. More precisely, the radiated EM field, by virtue of causality, admits the far-field representation (8.1) and hence its contribution from the 'sphere at infinity' is vanishingly small. Consequently, upon neglecting the contribution of the electric-current surface density on the circuit's PEC surfaces, the corresponding radiation characteristics follow (cf. (8.2))

$$\mathbf{I}_t \, \boldsymbol{E}_\infty^T(\boldsymbol{\xi}, t) = \frac{d}{c_0} \, \boldsymbol{\xi} \times \int_{\boldsymbol{x}' \in \partial\Omega} \boldsymbol{E}^T(\boldsymbol{x}', t + \boldsymbol{\xi} \cdot \boldsymbol{x}'/c_0) \times \boldsymbol{\nu}(\boldsymbol{x}') \mathrm{d}l(\boldsymbol{x}') \tag{10.5}$$

Equation (10.5) expresses the pulsed radiation characteristic of the electric-type in terms of the (equivalent) magnetic-current surface density on the circuit's periphery. More details about the radiation integral including its TD-CIM-based numerical calculation can be found in Chapter 8. Further applications concerning the TD radiated susceptibility of a planar circuit, with emphasis on the rectangular circuit, will be discussed in Chapter 12.

10.3 Receiving state of a planar circuit

The planar circuit is, in its receiving state, activated via a uniform TD plane wave (see Fig. 10.1) defined via

$$\boldsymbol{E}^i(\boldsymbol{x},t) = \boldsymbol{\alpha}\, e^i(t - \boldsymbol{\beta} \cdot \boldsymbol{x}/c_0) \tag{10.6}$$

$$\boldsymbol{H}^i(\boldsymbol{x},t) = (\epsilon_0/\mu_0)^{1/2}\boldsymbol{\beta} \times \boldsymbol{\alpha}\, e^i(t - \boldsymbol{\beta} \cdot \boldsymbol{x}/c_0) \tag{10.7}$$

where $e^i(t)$ is the plane-wave time signature, $\boldsymbol{\alpha}$ is the polarization vector and $\boldsymbol{\beta}$ denotes a unit vector in the direction of propagation. In the receiving situation, the presence of the planar circuit is accounted for by the scattered field (cf. Eq. (9.4))

$$\{\boldsymbol{E}^s, \boldsymbol{H}^s\} = \{\boldsymbol{E}^R - \boldsymbol{E}^i, \boldsymbol{H}^R - \boldsymbol{H}^i\} \tag{10.8}$$

where superscript R refers to the total field in the receiving situation. The scattered EM wave field is, in the (unbounded) domain exterior to the planar structure, source-free and satisfies

$$\boldsymbol{\nabla} \times \boldsymbol{H}^s - \epsilon_0 \partial_t \boldsymbol{E}^s = \boldsymbol{0} \tag{10.9}$$

$$\boldsymbol{\nabla} \times \boldsymbol{E}^s + \mu_0 \partial_t \boldsymbol{H}^s = \boldsymbol{0} \tag{10.10}$$

together with the 'radiation condition' over 'the bounding sphere at infinity', which is an attribute of its causal behavior.

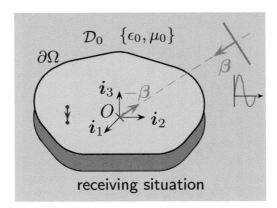

Figure 10.2: Planar circuit in its receiving state. © [2014] IEEE. Reprinted, with permission, from [125].

10.4 Reciprocity relations

Similarly to Sec. 9.3, the reciprocity theorem is applied in the first step to the domain exterior, to the circuit and to the transmitted (T) and scattered (s) wave fields. This step yields

$$\int_{\boldsymbol{x}\in\partial\mathcal{D}} \left[\boldsymbol{E}^T(\boldsymbol{x},t) \stackrel{*}{\times} \boldsymbol{H}^s(\boldsymbol{x},t)\right.$$
$$\left.- \boldsymbol{E}^s(\boldsymbol{x},t) \stackrel{*}{\times} \boldsymbol{H}^T(\boldsymbol{x},t)\right] \cdot \boldsymbol{\nu}(\boldsymbol{x})\mathrm{d}A(\boldsymbol{x}) = 0 \quad (10.11)$$

for all $t > 0$. Substitution of Eq. (10.8) in (10.11) gives

$$\int_{\boldsymbol{x}\in\partial\mathcal{D}} \left[\boldsymbol{E}^T(\boldsymbol{x},t) \stackrel{*}{\times} \boldsymbol{H}^R(\boldsymbol{x},t)\right.$$
$$\left.- \boldsymbol{E}^R(\boldsymbol{x},t) \stackrel{*}{\times} \boldsymbol{H}^T(\boldsymbol{x},t)\right] \cdot \boldsymbol{\nu}(\boldsymbol{x})\mathrm{d}A(\boldsymbol{x})$$
$$= \int_{\boldsymbol{x}\in\partial\mathcal{D}} \left[\boldsymbol{E}^T(\boldsymbol{x},t) \stackrel{*}{\times} \boldsymbol{H}^i(\boldsymbol{x},t)\right.$$
$$\left.- \boldsymbol{E}^i(\boldsymbol{x},t) \stackrel{*}{\times} \boldsymbol{H}^T(\boldsymbol{x},t)\right] \cdot \boldsymbol{\nu}(\boldsymbol{x})\mathrm{d}A(\boldsymbol{x}) \quad (10.12)$$

In the second step, the reciprocity theorem is applied to the domain occupied by the circuit, $\mathcal{D}$, and to the total wave fields in the both states, that is

$$\int_{\boldsymbol{x}\in\partial\mathcal{D}} \left[\boldsymbol{E}^T(\boldsymbol{x},t) \stackrel{*}{\times} \boldsymbol{H}^R(\boldsymbol{x},t) - \boldsymbol{E}^R(\boldsymbol{x},t) \stackrel{*}{\times} \boldsymbol{H}^T(\boldsymbol{x},t)\right] \cdot \boldsymbol{\nu}(\boldsymbol{x})\mathrm{d}A(\boldsymbol{x})$$
$$= \int_{\boldsymbol{x}\in\mathcal{D}} \boldsymbol{J}^T(\boldsymbol{x},t) \stackrel{*}{\cdot} \boldsymbol{E}^R(\boldsymbol{x},t)\mathrm{d}V(\boldsymbol{x})$$
$$(10.13)$$

In the final step, we combine (10.5) with (10.7) and (10.12) with (10.13), which finally yields

$$\boldsymbol{\alpha} \cdot \boldsymbol{E}^T_\infty(-\boldsymbol{\beta},t) \simeq \mathcal{V}^R(\boldsymbol{x}^S,t) \quad (10.14)$$

provided that

$$e^i(t) = \mu_0\partial_t\mathcal{I}^T(t) \quad (10.15)$$

where we have assumed the spatially concentrated vertical electric-current excitation port described via

$$\boldsymbol{J}^T(\boldsymbol{x},t) = \mathcal{I}^T(t)\delta(\boldsymbol{x} - \boldsymbol{x}^S)\,\boldsymbol{i}_3 \quad (10.16)$$

The final result (10.14) with (10.15) relates the pulsed EM radiation characteristics $\boldsymbol{E}_\infty^T$ of a planar circuit in the transmitting situation with the pulsed-voltage response $\mathcal{V}^R$ on the TD plane wave in its receiving state. Equations (10.14) and (10.15) thus provide an efficient means for determining the pulsed EM radiation characteristics of a thin planar antenna in the transmitting situation via its response to the plane wave in the receiving state (see Figs. 10.2 and 10.1). In fact, as will be shown in Chapter 11, the pulsed-voltage response can be interpreted as the (induced) voltage generator of the corresponding equivalent Thévenin's (one-port) circuit representation.

10.5   Numerical results

The derived reciprocity relation (10.14) with (10.15) is validated using FIT as implemented in CST Microwave Studio®. To this end we take four observation directions $\boldsymbol{\xi} = -\boldsymbol{\beta}$ such that the polarization and propagation vectors read

$$\boldsymbol{\alpha} = [0, \cos(\gamma), \sin(\gamma)] \tag{10.17}$$

$$\boldsymbol{\beta} = [0, -\sin(\gamma), \cos(\gamma)] \tag{10.18}$$

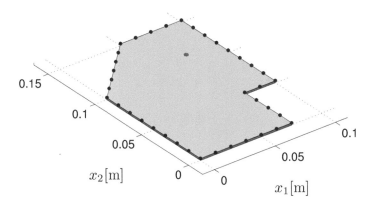

Figure 10.3: Computational model of the analysed circuit activated by a vertical electric-current port (the dot on the patch).

respectively, for $\gamma = \{3/4, 5/6, 11/12, 1\}\pi$. Note that $\boldsymbol{\alpha} \cdot \boldsymbol{\beta} = 0$. The calculations are carried out for the planar circuit shown in Fig. 10.3. The dielectric slab of the circuit has thickness $d = 1.50\,[\mathrm{mm}]$, electric

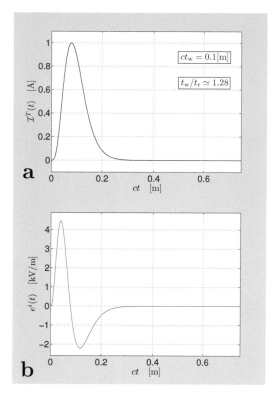

Figure 10.4: Excitation pulse shapes. (a) The power-exponential electric-current signature in the transmitting state; (b) the plane-wave signature in the receiving state.

permittivity $\epsilon = 4.0\epsilon_0$ and magnetic permeability $\mu = \mu_0$. The time window of observation is taken as $\{0 \leq c_0 t \leq 3.0\}$ [m].

The planar circuit is in its transmitting state excited via the vertical electric-current port with its center at $\{x_1^T, x_2^T\} = \{75.0, 112.5\}$ [mm]. As the corresponding electric-current excitation, we take the power-exponential pulse shape [96]

$$\mathcal{I}^T(t) = A(t/t_r)^\nu \exp[-\nu(t/t_r - 1)]\mathrm{H}(t) \qquad (10.19)$$

with $ct_w = 0.10$ [m], $\nu = 4$ and $A = 1.0$ [A]. The pulse time width t_w is then related to t_r and ν via $t_w = t_r \, \nu^{-\nu-1}\Gamma(\nu + 1) \exp(\nu)$ where $\Gamma(x)$ is the Euler gamma function. The corresponding electric-current excitation signature $\mathcal{I}^T$ for the transmitting state and the electric-field plane-wave signature e^i in the receiving state are shown in Fig. 10.4. In the transmitting situation, the analysed circuit is excited by the electric-current excitation pulse and the pulsed radiation field is observed us-

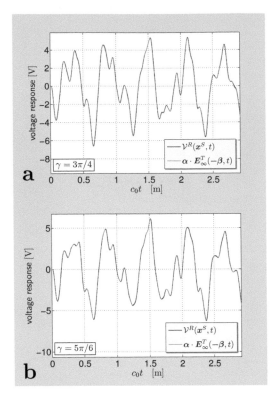

Figure 10.5: The pulsed voltage responses on the plane-wave excitation and the co-polarized pulsed radiation electric-field characteristics at $\boldsymbol{\xi} = -\boldsymbol{\beta}$ specified by (a) $\gamma = 3\pi/4$; (b) $\gamma = 5\pi/6$. © [2014] IEEE. Reprinted, with permission, from [125].

ing the 'far-field probes'. The obtained transmitted electric-field vector $\boldsymbol{E}_\infty^T(-\boldsymbol{\beta}, t)$ is then projected onto the polarization direction, specified by $\boldsymbol{\alpha}$. Subsequently, the TD voltage responses $\mathcal{V}^R$ on the corresponding pulsed plane waves are evaluated in the receiving state. The results are shown in Figs. 10.5 and 10.6. As can be observed, the pulse shapes evaluated in the transmitting and the receiving state overlap each other, thus validating the introduced (time-derivative) self-reciprocity relation (10.14)–(10.15). Since these pulses were evaluated via two distinct ways, we may conclude that the CST Microwave Studio® has passed the consistency check very well.

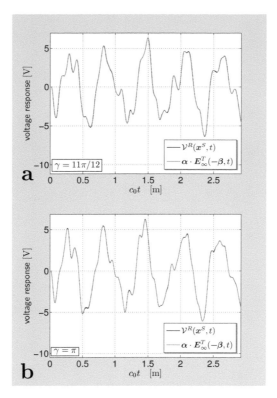

Figure 10.6: The pulsed voltage responses on the plane-wave excitation and the co-polarized pulsed radiation electric-field characteristics at $\boldsymbol{\xi} = -\boldsymbol{\beta}$ specified by (a) $\gamma = 11\pi/12$; (b) $\gamma = \pi$. © [2014] IEEE. Reprinted, with permission, from [125].

10.6 Conclusions

Self-reciprocity properties of a planar (one-port) circuit have been studied directly in TD. The derived transmission-reception 'time-derivative' self-reciprocity relation links the planar circuit's pulsed EM radiation characteristics to its pulsed-voltage response to a plane wave in the receiving situation. It has been demonstrated that the introduced relation is useful for consistency benchmarking of purely numerical EM solvers. The obtained results will be further generalized in the following Chapter 11, where the equivalent N-port Kirchhoff-network representation of a loaded planar circuit is studied in detail.

Chapter 11

Thévenin's Circuit of an N-port Planar Circuit

[1] The Kirchhoff-type equivalent network representation of an antenna system is a corollary of the Lorentz reciprocity theorem of the time-convolution type that illuminates its transmission/reception behavior and, thus, facilitates efficient antenna design and measurement methodologies. In the realm of EMC, the crucial objective is to secure the proper operation of an electronic device in the presence of an EM disturbance (i.e. the receiving situation) without introducing intolerable EM emissions (i.e the transmitting situation). Accordingly, the relation between the transmitting and receiving situations is of high importance in EMC, too (e.g. [86, Ch. 8]).

The transmitting situation of planar circuits is traditionally analysed with the aid of the cavity model that has proved to be computationally efficient and physically instructive in analyzing such structures of simple [59, 126] as well as irregular shapes [127, 136]. For the corresponding EM radiated susceptibility analysis concerning the plane-wave coupling into PCB traces, efficient computational models have been proposed and successfully validated (e.g. [60]). Despite the well-known benefits of utilizing the general property of (self-)reciprocity (e.g. [6]), it seems that its potentialities in analyzing radiated EM emissions/-susceptibility of planar circuits have not been fully appreciated so far. Apart from the reciprocity-based calculations in transmission-line the-

---

[1]© [2017] IEEE. This chapter is in part adapted, with permission, from [131].

ory (see e.g. [116, Sec. 7.5.2]), the vast majority of relevant works on this subject keep the transmission and reception situations apart. Hence, the main purpose of this chapter is to generalize the results of Chapter 10 and introduce a reciprocity-based description that will shed some light on the transmitting/receiving states of N-port planar circuits.

The sections of this chapter are organized in the following way. After introducing the problem configuration in Sec. 11.1, a reciprocity-based analysis of a receiving/transmitting N-port planar circuit is carried out. This analysis results in the N-port Thévenin-network representation of a planar circuit. Its Thévenin's open-circuit voltage is expressed for the TD incident EM plane wave as well as for the primary field generated by known EM-sources having their bounded support exterior to the planar structure. An application of the results is further illustrated on a 2-port planar circuit, whose equivalent circuit is discussed in detail. In this respect, it is demonstrated that the Kirchhoff-type equivalent circuit readily offers several ways for calculating the circuit's pulsed EM radiation characteristics. Finally, the obtained results are validated using FIT and the Feature Selective Validation (FSV) analysis [33, 84].

11.1 Model definition

We shall next analyse the pulsed EM properties of a multiport planar circuit that occupies a bounded domain $\mathcal{D} \subset \mathbb{R}^3$ enclosed by the surface $\partial \mathcal{D} \subset \mathbb{R}^2$. The planar circuit is placed in the linear, isotropic, homogeneous and loss-free embedding $\mathcal{D}_0$ whose EM properties are described by its electric permittivity ϵ_0 and magnetic permeability μ_0. The corresponding EM wave speed is $c_0 = (\epsilon_0 \mu_0)^{-1/2} > 0$. The EM properties of the dielectric slab are described by its electric permittivity ϵ and magnetic permeability μ_0. The electrically conducting plate of the circuit occupies a surface domain $\Omega \subset \partial \mathcal{D}$ bounded by its rim $\partial \Omega \subset \mathbb{R}$. The normal outer (unit) vector is denoted by $\boldsymbol{\nu}$. In the following sections we will study TD reciprocity relations between transmitting and receiving behavior of an N-port planar circuit.

11.2 Transmitting state of an N-port planar circuit

The planar circuit is in the transmitting state (see Fig. 11.1) activated by vertical electric-current ports whose action is accounted for by the electric-current volume density $\boldsymbol{J}^T(\boldsymbol{x}|\boldsymbol{x}^m, t)$, for $m = \{1, \ldots, N\}$, where

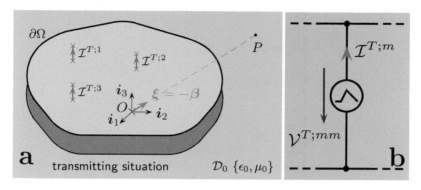

Figure 11.1: Transmitting situation. (a) Emitting N-port planar circuit; (b) a feeding port. © [2017] IEEE. Adapted, with permission, from [131].

$\boldsymbol{x}^m$ is the position of the m-th port and $\mathcal{I}^{T;m}$ denotes the corresponding electric-current excitation pulse shape. The pulsed voltage response $\mathcal{V}^{T;nm}$ at $\boldsymbol{x}^n$ is then linearly related to the excitation current at $\boldsymbol{x}^m$ via

$$\mathcal{V}^{T;nm}(t) = \mathcal{Z}^{T;nm}(t) * \mathcal{I}^{T;m}(t) \tag{11.1}$$

for $n = \{1, \dots, N\}$, where $\mathcal{Z}^{T;nm}$ is the corresponding transfer impedance that for $m = n$ can be interpreted as the input impedance. Consequently, the planar circuit emits the EM wave field $\{\boldsymbol{E}^T, \boldsymbol{H}^T\}(\boldsymbol{x}, t)$ into its embedding, where this radiated field admits the far-field expansion [24, Sec. 26.12]

$$\{\boldsymbol{E}^T, \boldsymbol{H}^T\}(\boldsymbol{x}, t) = \sum_{m=1}^{N} \frac{\{\boldsymbol{E}_\infty^T, \boldsymbol{H}_\infty^T\}(\boldsymbol{\xi}, t - |\boldsymbol{x}|/c_0)|_{\mathcal{I}^{T;m}(t)=\delta(t)}}{4\pi|\boldsymbol{x}|} * \mathcal{I}^{T;m}(t)$$
$$[1 + \mathcal{O}(|\boldsymbol{x}|^{-1})] \tag{11.2}$$

as $|\boldsymbol{x}| \to \infty$, where $\{\boldsymbol{E}_\infty^T, \boldsymbol{H}_\infty^T\} = \{\boldsymbol{E}_\infty^T, \boldsymbol{H}_\infty^T\}(\boldsymbol{\xi}, t)$ are the electric- and magnetic-field vector radiation characteristics, respectively, and $\boldsymbol{\xi} = \boldsymbol{x}/|\boldsymbol{x}|$ is the unit vector in the direction of observation. The TD surface-source radiation-characteristics representation follows as [125, Eq. (8)]

$$\mathbf{I}_t \boldsymbol{E}_\infty^T(\boldsymbol{\xi}, t) = \mu_0 \boldsymbol{\xi} \times \boldsymbol{\xi} \times \int_{\boldsymbol{x} \in \partial \mathcal{D}} \boldsymbol{\nu}(\boldsymbol{x}) \times \boldsymbol{H}^T(\boldsymbol{x}, t + \boldsymbol{\xi} \cdot \boldsymbol{x}/c_0) \mathrm{d}A(\boldsymbol{x})$$
$$- c_0^{-1} \boldsymbol{\xi} \times \int_{\boldsymbol{x} \in \partial \mathcal{D}} \boldsymbol{\nu}(\boldsymbol{x}) \times \boldsymbol{E}^T(\boldsymbol{x}, t + \boldsymbol{\xi} \cdot \boldsymbol{x}/c_0) \mathrm{d}A(\boldsymbol{x}) \tag{11.3}$$

with $\boldsymbol{\xi} \times \boldsymbol{E}_\infty^T = (\mu_0/\epsilon_0)^{1/2} \boldsymbol{H}_\infty^T$ and $\boldsymbol{\xi} \cdot \boldsymbol{E}_\infty^T = 0$.

11.3 Receiving states of an N-port planar circuit

In reception, we shall distinguish between two different scenarios. In the first one, the planar circuit is, similarly to Sec. 10.3, irradiated by a uniform EM plane wave defined in Eqs. (10.6) and (10.7) (see Fig. 11.2a). Consequently, the total field in the configuration, $\{E^R, H^R\}(x, t)$, is the superposition of the incident field $\{E^i, H^i\}(x, t)$ and the scattered field $\{E^s, H^s\}(x, t)$ that is defined according to Eq. (10.8). In the second scenario, the incident field is generated by the known electric- and magnetic-current sources $\{J^R, K^R\}$ that are distributed exterior to the planar circuit and whose support is a bounded domain $\mathcal{D}^R$ (see Fig. 11.2b).

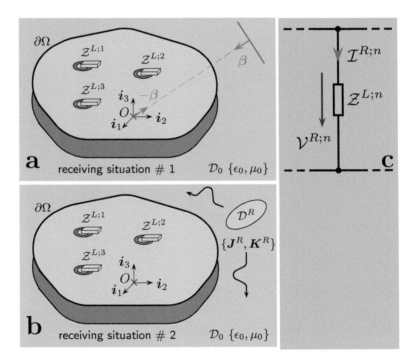

Figure 11.2: Receiving situations where the receiving N-port planar circuit is irradiated by (a) the EM plane wave or (b) the known EM source distribution; (c) a load impedance. © [2017] IEEE. Adapted, with permission, from [131].

The receiving planar structure is at x^n, for $n = \{1, \dots, N\}$, loaded by lumped impedances. The corresponding electric-current volume density is then described via $J^R(x, t)$, with $\mathcal{I}^{R;n}$ being the (induced) electric current flowing across the n-th load. The voltage across the load is then linearly related to the (linear) load current according to

$$\mathcal{V}^{R;n}(t) = \mathcal{Z}^{L;n}(t) * \mathcal{I}^{R;n}(t) \tag{11.4}$$

where $\mathcal{Z}^{L;n}$ is the n-th load's impedance (see Fig. 11.2c).

11.4 Reciprocity analysis for the incident plane wave

The present reciprocity analysis refers to the receiving situation depicted in Fig. 11.2a and partially overlaps with the one given in Sec. 10.4. For convenience of the reader, however, we shall next provide a detailed step-by-step derivation. To that end we first combine the surface-integral representation of the transmitted-field radiation amplitude (11.3) with Eqs. (10.6)–(10.7) and get the following integral relation

$$\int_{\boldsymbol{x}\in\partial\mathcal{D}} \left[\boldsymbol{E}^T(\boldsymbol{x},t) \overset{*}{\times} \boldsymbol{H}^i(\boldsymbol{x},t) - \boldsymbol{E}^i(\boldsymbol{x},t) \overset{*}{\times} \boldsymbol{H}^T(\boldsymbol{x},t) \right] \cdot \boldsymbol{\nu}(\boldsymbol{x})\mathrm{d}A(\boldsymbol{x})$$

$$= -\mu_0^{-1}\boldsymbol{\alpha}e^i(t) \overset{*}{\cdot} \mathbf{I}_t \boldsymbol{E}_\infty^T(-\boldsymbol{\beta},t) \tag{11.5}$$

Owing to the fact that the transmitted and scattered wave EM fields are causal and that the embedding $\mathcal{D}_0$ is self-adjoint in its EM constitutive properties, we can write

$$\int_{\boldsymbol{x}\in\partial\mathcal{D}} \left[\boldsymbol{E}^T(\boldsymbol{x},t) \overset{*}{\times} \boldsymbol{H}^s(\boldsymbol{x},t) \right.$$

$$\left. - \boldsymbol{E}^s(\boldsymbol{x},t) \overset{*}{\times} \boldsymbol{H}^T(\boldsymbol{x},t) \right] \cdot \boldsymbol{\nu}(\boldsymbol{x})\mathrm{d}A(\boldsymbol{x}) = 0 \tag{11.6}$$

which in combination with (11.5) and (10.8) yields

$$\int_{\boldsymbol{x}\in\partial\mathcal{D}} \left[\boldsymbol{E}^T(\boldsymbol{x},t) \overset{*}{\times} \boldsymbol{H}^R(\boldsymbol{x},t) - \boldsymbol{E}^R(\boldsymbol{x},t) \overset{*}{\times} \boldsymbol{H}^T(\boldsymbol{x},t) \right] \cdot \boldsymbol{\nu}(\boldsymbol{x})\mathrm{d}A(\boldsymbol{x})$$

$$= -\mu_0^{-1}\boldsymbol{\alpha}e^i(t) \overset{*}{\cdot} \mathbf{I}_t \boldsymbol{E}_\infty^T(-\boldsymbol{\beta},t) \tag{11.7}$$

Application of the reciprocity theorem to the domain occupied by the planar circuit and to the total fields in the transmitting and receiving situations yields

$$\int_{\boldsymbol{x}\in\partial\mathcal{D}} \left[\boldsymbol{E}^T(\boldsymbol{x},t) \overset{*}{\times} \boldsymbol{H}^R(\boldsymbol{x},t) - \boldsymbol{E}^R(\boldsymbol{x},t) \overset{*}{\times} \boldsymbol{H}^T(\boldsymbol{x},t) \right] \cdot \boldsymbol{\nu}(\boldsymbol{x})\mathrm{d}A(\boldsymbol{x})$$

$$= \int_{\boldsymbol{x}\in\mathcal{D}} \left[\boldsymbol{J}^T(\boldsymbol{x},t) \overset{*}{\cdot} \boldsymbol{E}^R(\boldsymbol{x},t) - \boldsymbol{J}^R(\boldsymbol{x},t) \overset{*}{\cdot} \boldsymbol{E}^T(\boldsymbol{x},t) \right] \mathrm{d}V(\boldsymbol{x}) \tag{11.8}$$

where we have assumed that the medium in $\mathcal{D}$ is self-adjoint in its EM properties. Subsequently, combination of (11.7) with (11.8) leads to

$$\int_{\boldsymbol{x} \in \mathcal{D}} \left[\boldsymbol{J}^T(\boldsymbol{x}, t) \overset{*}{\cdot} \boldsymbol{E}^R(\boldsymbol{x}, t) - \boldsymbol{J}^R(\boldsymbol{x}, t) \overset{*}{\cdot} \boldsymbol{E}^T(\boldsymbol{x}, t) \right] dV(\boldsymbol{x})$$

$$= -\mu_0^{-1} \alpha e^i(t) \overset{*}{\cdot} \mathbf{I}_t \boldsymbol{E}_\infty^T(-\boldsymbol{\beta}, t) \qquad (11.9)$$

By virtue of the thin-slab assumption, the latter relation can be further rewritten as

$$\mathcal{I}^{T;m}(t) * \mathcal{V}^{R;m}(t) + \sum_{n=1}^{N} \mathcal{I}^{R;n}(t) * \mathcal{V}^{T;nm}(t)$$

$$= \mu_0^{-1} \alpha e^i(t) \overset{*}{\cdot} \mathbf{I}_t \boldsymbol{E}_\infty^T(-\boldsymbol{\beta}, t) \qquad (11.10)$$

for $m = \{1, \ldots, N\}$, where the orientation of the source/load currents and voltages is depicted in Figs. 11.1b and 11.2c. Equation (11.10) makes it possible to introduce both Norton's and Thévenin's equivalent networks of an N-port planar circuit. As these equivalent circuits are fully equivalent (see [29]), we shall, without loss of generality, further limit our analysis to the Thévenin network only. To this end, we assume that the planar circuit is in its transmitting situation excited by prescribed electric-current source signatures $\mathcal{I}^{T;m}$ and rewrite Eq. (11.10) as

$$\mathcal{V}^{R;m}(t) + \sum_{n=1}^{N} \mathcal{Z}^{T;nm}(t) * \mathcal{I}^{R;n}(t) = \mathcal{V}^{G;m}(t) \qquad (11.11)$$

with

$$\mathcal{V}^{G;m}(t) = \mu_0^{-1} \alpha e^i(t) \overset{*}{\cdot} \mathbf{I}_t \boldsymbol{E}_\infty^T(-\boldsymbol{\beta}, t)|_{\mathcal{I}^{T;m}(t) = \delta(t)} \qquad (11.12)$$

for $m = \{1, \ldots, N\}$, being the Thévenin voltage generator. Equations (11.11) and (11.12) define the Thévenin network representation of an N-port planar circuit. The conditions under which the impedance matrix $\mathcal{Z}^T$ is symmetrical have been discussed in [131].

11.5 Reciprocity analysis for the incident wave field generated by known sources

The reciprocity analysis that follows refers to the receiving scenario shown in Fig. 11.2b. In such a case, the reciprocity theorem is first applied to the domain exterior to the circuit and to the total fields in the transmitting and receiving states. Considering the orientation of the outer normal vector $\boldsymbol{\nu}$, we get

$$\int_{\boldsymbol{x}\in\partial\mathcal{D}} \left[\boldsymbol{E}^T(\boldsymbol{x},t) \overset{*}{\times} \boldsymbol{H}^R(\boldsymbol{x},t) - \boldsymbol{E}^R(\boldsymbol{x},t) \overset{*}{\times} \boldsymbol{H}^T(\boldsymbol{x},t)\right] \cdot \boldsymbol{\nu}(\boldsymbol{x}) \mathrm{d}A(\boldsymbol{x})$$

$$= \int_{\boldsymbol{x}\in\mathcal{D}^R} \left[\boldsymbol{J}^R(\boldsymbol{x},t) \overset{*}{\cdot} \boldsymbol{E}^T(\boldsymbol{x},t) - \boldsymbol{K}^R(\boldsymbol{x},t) \overset{*}{\cdot} \boldsymbol{H}^T(\boldsymbol{x},t)\right]\mathrm{d}V(\boldsymbol{x}) \quad (11.13)$$

As in the previous section, the TD reciprocity theorem of the time-convolution type is next applied to the total field states and to the (bounded) domain occupied by the planar circuit, which yields Eq. (11.8). Upon combining the latter with Eq. (11.13), we may, under the thin-slab approximation, arrive at the following relation (cf. Eq. (11.10))

$$\mathcal{I}^{T;m}(t) * \mathcal{V}^{R;m}(t) + \sum_{n=1}^{N} \mathcal{I}^{R;n}(t) * \mathcal{V}^{T;nm}(t)$$

$$= \int_{\boldsymbol{x}\in\mathcal{D}^R} \left[\boldsymbol{K}^R(\boldsymbol{x},t) \overset{*}{\cdot} \boldsymbol{H}^T(\boldsymbol{x},t) - \boldsymbol{J}^R(\boldsymbol{x},t) \overset{*}{\cdot} \boldsymbol{E}^T(\boldsymbol{x},t)\right]\mathrm{d}V(\boldsymbol{x}) \quad (11.14)$$

for $m = \{1,\dots,N\}$, that readily yields both (equivalent) Kirchhoff-network representations. Choosing, again, the Thévenin representation described via Eq. (11.11), its voltage generator for the incident wave field generated by known source distributions $\boldsymbol{J}^R, \boldsymbol{K}^R$ is written as (cf. Eq. (11.12))

$$\mathcal{V}^{G;m}(t) = \int_{\boldsymbol{x}\in\mathcal{D}^R} \left[\boldsymbol{K}^R(\boldsymbol{x},t) \overset{*}{\cdot} \boldsymbol{H}^T(\boldsymbol{x},t)|_{\mathcal{I}^{T;m}(t)=\delta(t)}\right.$$

$$\left. - \boldsymbol{J}^R(\boldsymbol{x},t) \overset{*}{\cdot} \boldsymbol{E}^T(\boldsymbol{x},t)|_{\mathcal{I}^{T;m}(t)=\delta(t)}\right]\mathrm{d}V(\boldsymbol{x}) \quad (11.15)$$

for $m = \{1,\dots,N\}$, which completes our TD reciprocity-based description of an N-port planar circuit.

11.6   An illustrative example

As has been shown in [131], the main results of the previous Sec. 11.4 can be illustrated clearly on the evaluation of the pulsed EM radiation from a 2-port planar circuit. This procedure is, hence, followed next. For a 2-port circuit, the relevant equations follow directly from Eq. (11.11) with $N = 2$, viz

$$\begin{pmatrix} \mathcal{V}^{G;1} \\ \mathcal{V}^{G;2} \end{pmatrix} = \begin{pmatrix} \mathcal{V}^{R;1} \\ \mathcal{V}^{R;2} \end{pmatrix} + \begin{pmatrix} \mathcal{Z}^{T;11} & \mathcal{Z}^{T;21} \\ \mathcal{Z}^{T;12} & \mathcal{Z}^{T;22} \end{pmatrix} * \begin{pmatrix} \mathcal{I}^{R;1} \\ \mathcal{I}^{R;2} \end{pmatrix} \quad (11.16)$$

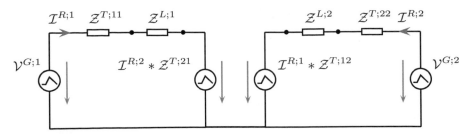

Figure 11.3: Equivalent circuit of a 2-port planar circuit. © [2017] IEEE. Adapted, with permission, from [131].

The corresponding network representation of the 2-port planar circuit is shown in Fig. 11.3.

It is further observed that one can decouple the system of equations (11.16) by considering its special cases, namely, the open-circuited port 2 for which $\mathcal{I}^{R;2} = 0$, i.e.

$$\mathcal{V}^{R;1}(t) + \mathcal{Z}^{T;11}(t) * \mathcal{I}^{R;1}(t) = \mathcal{V}^{G;1}(t) \tag{11.17}$$

$$\mathcal{V}^{R;2}(t) + \mathcal{Z}^{T;12}(t) * \mathcal{I}^{R;1}(t) = \mathcal{V}^{G;2}(t) \tag{11.18}$$

and, the case when port 1 is open-circuited, i.e. $\mathcal{I}^{R;1} = 0$, which implies

$$\mathcal{V}^{R;1}(t) + \mathcal{Z}^{T;21}(t) * \mathcal{I}^{R;2}(t) = \mathcal{V}^{G;1}(t) \tag{11.19}$$

$$\mathcal{V}^{R;2}(t) + \mathcal{Z}^{T;22}(t) * \mathcal{I}^{R;2}(t) = \mathcal{V}^{G;2}(t) \tag{11.20}$$

Now, using Eq. (11.12), it is obvious that we may use either Eq. (11.17) or (11.19) to calculate the pulsed EM characteristics due to the electric-current source of port 1, i.e. $\mathcal{I}^{T;1}$, while the use of either Eq. (11.18) or (11.20) leads to the pulsed EM characteristics due to $\mathcal{I}^{T;2}$. In the numerical examples that follow we shall consider the former case and rewrite the corresponding equations to the form that is practical for our calculations, namely

$$\mathcal{I}^{T;1}(t) * \mathcal{V}^{R;1}(t) + \mathcal{V}^{T;11}(t) * \mathcal{I}^{R;1}(t)$$
$$= \mathcal{I}^{T;1}(t) * \mathcal{V}^{G;1}(t) \text{ for } \mathcal{I}^{R;2}(t) = 0 \tag{11.21}$$

$$\mathcal{I}^{T;1}(t) * \mathcal{V}^{R;1}(t) + \mathcal{V}^{T;21}(t) * \mathcal{I}^{R;2}(t)$$
$$= \mathcal{I}^{T;1}(t) * \mathcal{V}^{G;1}(t) \text{ for } \mathcal{I}^{R;1}(t) = 0 \tag{11.22}$$

The procedure then goes along the following lines

- For a given excitation pulse $\mathcal{I}^{T;1}$, calculate (or measure) the pulsed voltage responses $\mathcal{V}^{T;11}$ or $\mathcal{V}^{T;21}$ in the transmitting state. Such calculations can be readily carried out using TD-CIM.

- For the corresponding plane-wave pulse e^i (see Eq. (11.24)), calculate (or measure) the pulsed voltage or/and current responses $\{\mathcal{V}^{R;1}, \mathcal{I}^{R;1}\}$ with $\mathcal{I}^{R;2} = 0$ or $\{\mathcal{V}^{R;1}, \mathcal{I}^{R;2}\}$ with $\mathcal{I}^{R;1} = 0$.

- With the pulsed responses at our disposal, we can perform the operations indicated in Eq. (11.21) or (11.22) and get $\mathcal{I}^{T;1} * \mathcal{V}^{G;1}$. Recovery of $\mathcal{V}^{G;1}$ calls for a deconvolution algorithm. An example of the latter is specified in the following Sec. 11.7.

- The Thévenin voltage generator $\mathcal{V}^{G;1}$ is subsequently identified with the corresponding TD EM radiation characteristics via (cf. Eq. (11.12))

$$\mathcal{V}^{G;1}(t) = \boldsymbol{\alpha} \cdot \boldsymbol{E}_\infty^T(-\boldsymbol{\beta}, t)|_{\mathcal{I}^{T;1}(t)=\delta(t)} * \mathcal{I}^{T;1}(t) \qquad (11.23)$$

provided that the plane-wave signature is related to the electric-current pulse shape in the following way

$$e^i(t) = \mu_0 \partial_t \mathcal{I}^{T;1}(t) \qquad (11.24)$$

Finally note that Eqs. (11.23)–(11.24) are known as the transmission-reception time-derivative relation [125, Sec. IVb].

11.7   Numerical results

The proposed methodology will be validated by analyzing the irregularly-shaped 2-port planar circuit shown in Fig. 11.4. A similar structure has been previously analysed in [131], where the observed late-time discrepancies have been attributed to the problem formulation that depends upon the thin-slab approximation and its consequences (see Sec. 2.1). In order to examine this explanation, the thickness of the analysed planar structure has been reduced to $d = 0.75$ [mm]. The circuit has its accessible ports placed at $\boldsymbol{x}^1 = \{75.0, 112.5, 0\}$ [mm] (**PORT 1**) and at $\boldsymbol{x}^2 = \{25.0, 37.5, 0\}$ [mm] (**PORT 2**). The relative electric permittivity of the dielectric slab is $\epsilon_r = 4.2$. The corresponding EM wave speed in the slab is $c = c_0/\sqrt{\epsilon_r}$. For validation purposes, the transmitting state is analysed using TD-CIM, while the pulsed responses in the receiving situation are evaluated with the help of FIT of CST Microwave Studio®.

In accordance with the methodology summarized in Sec. 11.6, the planar circuit is first analysed in its transmitting state. To this end, the emitting planar structure is activated at its **PORT 1** by the electric-current pulse defined in Appendix F with the amplitude $A = 1.0\,[\text{A}]$ and the spatial pulse width $ct_{\text{w}} = 0.20\,[\text{m}]$ (see Fig. 11.5a). Note that the thickness of the analysed structure is very thin with respect to the spatial support of the pulse, $ct_{\text{w}}/d = 800/3$, thereby satisfying the condition for the thin-slab approximation. Figures 11.6a and 11.6b show the resulting pulsed-voltage responses $\mathcal{V}^{T;11}$ and $\mathcal{V}^{T;21}$ at **PORT 1** and **PORT 2**, respectively. In the subsequent step, the analysed structure is irradiated by the incident uniform plane-wave (see Eqs. (10.6) and (10.7)) whose polarization and direction of propagation is defined according to Eqs. (10.17) and (10.18). In the latter we use $\gamma = 5\pi/6$. The plane-wave pulse shape calculated from Eq. (11.24) is shown in Fig. 11.5b. According to Eqs. (11.21) and (11.22), we will distinguish between two receiving scenarios. In the first one pertaining to Eq. (11.21), **PORT 2** is left open-circuited (i.e. $\mathcal{I}^{R;2} = 0$) and we calculate the pulsed response $\mathcal{V}^{R;1}$ at **PORT 1**. For this case we have chosen a purely resistive (instantaneously-reacting) load defined as $\mathcal{Z}^{L;1}(t) = R^{L;1}\delta(t)$, with $R^{L;1} = 240.0\,[\Omega]$, and calculated the pulsed-voltage response across the lumped element. This result is plotted in Fig. 11.7a. Since the corresponding electric-current response $\mathcal{I}^{R;1}$ is merely a scaled copy of $\mathcal{V}^{R;1}$, its plot is omitted here.

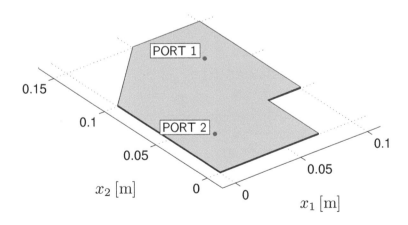

Figure 11.4: Computational model of the analysed circuit with two accessible ports (the dots on the patch). © [2017] IEEE. Adapted, with permission, from [131].

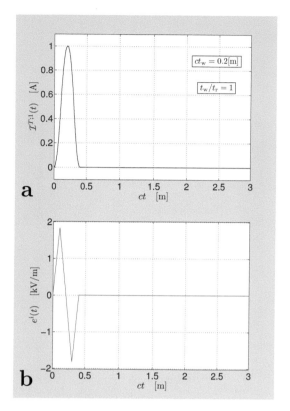

Figure 11.5: Excitation pulse shapes. (a) The bell-shaped electric-current signature in the transmitting state; (b) the plane-wave signature in the receiving state.

The second receiving state is associated with Eq. (11.22) and corresponds to the situation with the open-circuited **PORT 1** (i.e. $\mathcal{I}^{R;1} = 0$). In this case, we have calculated the open-circuited voltage response $\mathcal{V}^{R;1}$ and the electric-current pulse $\mathcal{I}^{R;2}$ flowing across the chosen resistive impedance $\mathcal{Z}^{L;2}(t) = R^{L;2}\delta(t)$ connected at **PORT 2** with $R^{L;2} = 50\,[\Omega]$. The resulting pulse shapes are shown in Figs. 11.7b and 11.7c, respectively.

With all the signals at our disposal, we may evaluate the time convolutions on the left-hand sides of Eqs. (11.21) and (11.22) and get $\mathcal{Q}(t) = \mathcal{I}^{T;1}(t) * \mathcal{V}^{G;1}(t)$. For the electric-current excitation pulse specified in Appendix F, the closed-form deconvolution algorithm that yields $\mathcal{V}^{G;1}$ at once exists, viz

$$\mathcal{V}^{G;1}(t) = \frac{t_{\mathrm{w}}^2}{4A} \sum_{n=0}^{\infty} \frac{2(n+2)^2 - 1 + (-1)^{n+2}}{8} \partial_t^3 \mathcal{Q}(t - nt_{\mathrm{w}}/2) \quad (11.25)$$

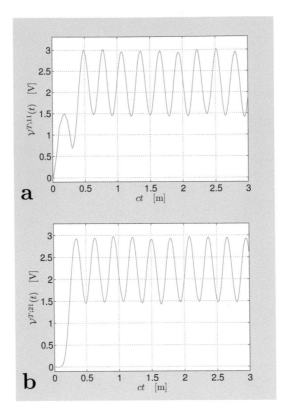

Figure 11.6: The pulsed voltage responses in the transmitting situation that are observed at (a) PORT 1; (b) PORT 2.

Obviously, the number of terms in (11.25) is finite in any (bounded) time-window of observation. For an alternative closed-form deconvolution algorithm we refer the reader to Eq. (9.17).

The sought pulse shapes of the 'co-polarized' pulsed EM radiation characteristics at $\xi = -\beta$ as calculated from Eqs. (11.21)–(11.22) are shown in Fig. 11.8a. To further validate the results, this radiated electric-field component has been evaluated with the help of the (referential) 'far-field probes' as implemented in CST Microwave Studio®. As can be seen, the final pulses are on top of each other. In order to assess the comparison, the FSV analysis has been carried out. Examples of the amplitude difference measure (ADM) and the feature difference measure (FDM) confidence histograms are given in Figs. 11.8b and 11.8c. Here, EX, VG, G, F, P, and VP stand for excellent, very good, good, fair, poor, and very poor, respectively (see [33, 84] for further details). In summary, the obtained results have confirmed the conclusions drawn in [131].

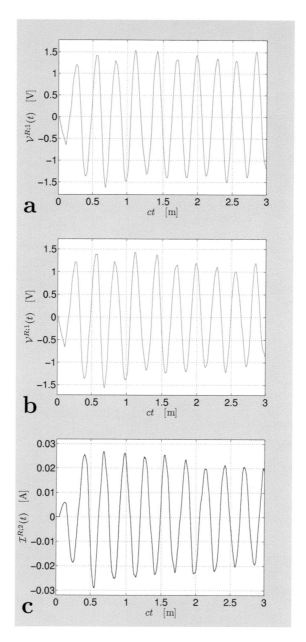

Figure 11.7: The pulsed responses in the receiving situation. (a) Voltage across $R^{L;1}$ at PORT 1 with PORT 2 open-circuited; (b) open-circuited voltage at PORT 1 with $R^{L;2}$ connected at PORT 2; (c) electric current across $R^{L;2}$ at PORT 2 with PORT 1 open-circuited.

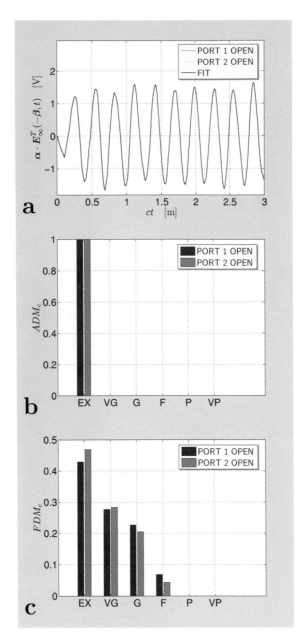

Figure 11.8: (a) Co-polarized pulsed EM radiation characteristics in $\boldsymbol{\xi} = -\boldsymbol{\beta}$ direction as evaluated via Eq. (11.21) (**PORT 2** open-circuited), Eq. (11.22) (**PORT 1** open-circuited) and the 'far-field probes' of the referential FIT; FSV analysis and its (b) *ADM* and (c) *FDM* histograms.

11.8   Conclusions

The Kirchhoff-type equivalent network representation of a planar circuit with N accessible ports has been constructed with the aid of the TD reciprocity theorem of the time-convolution type. The Thévenin voltage generator has been specified for the pulsed incident EM plane wave, as well as for known EM sources distributed in a bounded domain localized exterior to the planar circuit. The former case has been subsequently applied to closely analyse the TD emission/reception reciprocity properties of a 2-port planar circuit. The analysis has revealed two possible ways for calculating the pulsed EM radiation characteristics from the structure's response to the incident plane wave. In order to validate the TD reciprocity-based methodology, numerical experiments have been conducted. The consistency of the results has been further assessed with the aid of (three-dimensional) FIT and the FSV analysis.

Chapter 12

Time-domain Radiated Susceptibility of a Planar Circuit

[1] The cavity model, as formulated in Sec. 2.1, proved to be physically intuitive and computationally efficient for analyzing the pulsed EM radiation characteristics of planar circuits (see Chapter 8). Based on reciprocity considerations, it may be anticipated that this 2D model will do well in the corresponding receiving situation, in which the planar circuit is irradiated by an external EM source. The confirmation of this expectation is basically a spin-off of the present chapter, where straightforward expressions for the efficient calculation of the pulsed-voltage response to an external EM disturbance are constructed.

In this chapter, the reciprocity theorem of the time-convolution type is applied to express the pulsed voltage induced within a planar circuit via a straightforward one-dimensional integration of the testing (or auxiliary) field distribution along the circuit's rim. It is demonstrated that this approach is particularly efficient when combined with TD-CIM, whose basic formulation was introduced in Sec. 3, or the TD ray-type Green's function representation as given in Sec. 5. While the former leads directly to the required space-time testing voltage distribution on the circuit's rim, the latter expresses the solution in terms of (a finite number of) the causality-preserving TD ray-like constituents. In

[1] © [2016] Taylor & Francis Ltd. This chapter is largely based, with permission, on [130].

this way, the introduced methodology makes it possible to readily evaluate the TD radiated susceptibility of a planar circuit with very low computational efforts and yet reasonable accuracy.

Closely related scattering problems are dominantly solved using the standard real-FD numerical techniques. Examples in this category are Method-of-Moments-based solutions [74, 90] or the Finite-Element-based analysis concerning a microstrip structure and related topologies (see [12, 138], for example). As to EM plane-wave coupling, the previous works on the subject analyse the aperture coupling to a perfectly-conducting box [78, 106, 110] and approximate models of a PCB trace [8, 60], for instance, relying on the classical transmission-line theory [116, Chapter 7]. The theoretical developments presented in this chapter rely on the combination of the TD reciprocity theorem of the time-convolution type (see [24, Sec. 28.2] and [133, Sec. 1.4.1]) with the cavity-model based methodologies introduced in Chapters 3, 6 and 5.

12.1 Reciprocity relations

We consider a planar circuit that operates in the receiving state when it is irradiated by an impulsive plane wave defined via Eqs. (10.6) and (10.7) (see Fig. 10.2). For the sake of simplicity, the receiving circuit is again placed in the linear, isotropic, homogeneous and loss-free embedding $\mathcal{D}_0$ whose EM properties are described by its electric permittivity ϵ_0 and magnetic permeability μ_0. The corresponding EM wave speed is $c_0 = (\epsilon_0\mu_0)^{-1/2} > 0$. The EM properties of the dielectric slab are described by its electric permittivity ϵ and magnetic permeability μ_0. The electrically conducting plate of the circuit occupies a surface domain $\Omega^R \subset \partial\mathcal{D}^R$ bounded by its rim $\partial\Omega^R \subset \mathbb{R}$. The normal outer unit vector is denoted by $\boldsymbol{\nu}$, while the corresponding tangent vector is $\boldsymbol{\tau} = \boldsymbol{i}_3 \times \boldsymbol{\nu}$.

In order to describe the TD radiated susceptibility in closed form, we follow (in part) the procedure described in Sec. 9.3. Upon enforcing the magnetic-wall boundary condition for the testing field state, Eq. (9.7) can be, under the thin-slab approximation, rewritten as

$$\int_{\boldsymbol{x}'\in\partial\Omega^R} \mathcal{V}^B(\boldsymbol{x}'|\boldsymbol{x}^S,t) * \boldsymbol{\tau}(\boldsymbol{x}') \cdot \boldsymbol{H}^R(\boldsymbol{x}',t)\mathrm{d}l(\boldsymbol{x}')$$

$$= \int_{\boldsymbol{x}'\in\partial\Omega^R} \mathcal{V}^B(\boldsymbol{x}'|\boldsymbol{x}^S,t) * \boldsymbol{\tau}(\boldsymbol{x}') \cdot \boldsymbol{H}^i(\boldsymbol{x}',t)\mathrm{d}l(\boldsymbol{x}') \qquad (12.1)$$

for all $t > 0$, where $\mathcal{V}^B(\boldsymbol{x}|\boldsymbol{x}^S,t)$ denotes the testing voltage distribution along the rim of the receiving planar circuit $\partial\Omega^R$. It is worth noting that the tangential part (with respect to $\partial\Omega^R$) of the (total) field $\boldsymbol{H}^R$ in Eq. (12.1) should also be, in virtue of the thin-slab approximation, taken as zero. As the TD optical (extinction) theorem [135] tells, however, the cavity model in such a case does not absorb any energy. The boundary perturbation that allows for a non-vanishing tangential part of H^R is also commonly used in the corresponding transmitting situation (see [109, Sec. 4.7] and [41, Sec. 2.3], for example).

In the second step, the reciprocity theorem is applied to the domain occupied by the receiving planar circuit and to the total-field (R) and the testing-field (B) states, which, under the thin-slab approximation, yields (cf. Eq. (10.13))

$$\int_{x' \in \partial \Omega^R} \mathcal{V}^B(x'|x^S, t) * \tau(x') \cdot H^R(x', t) \mathrm{d}l(x')$$
$$= -\mathcal{V}^R(x^S, t) * \mathcal{I}^B(t) \qquad (12.2)$$

where $\mathcal{V}^R = -dE_3^R$ is the total voltage induced in the planar circuit at $x^S \in \Omega$. Upon combining the latter with Eq. (12.1) we end up with (cf. Eq. (9.9))

$$\mathcal{V}^R(x^S, t) * \mathcal{I}^B(t) =$$
$$= -\int_{x' \in \partial \Omega^R} \mathcal{V}^B(x'|x^S, t) * \tau(x') \cdot H^i(x', t) \mathrm{d}l(x') \qquad (12.3)$$

which expresses the induced (total) voltage $\mathcal{V}^R$ in the planar circuit though the testing voltage distribution $\mathcal{V}^B$ and the tangential part of the incident plane wave H^i along the circuit rim $\partial \Omega^R$. Now, taking into the account that Eq. (12.3) should hold for arbitrary $\mathcal{I}^B(t)$, we arrive at

$$\mathcal{V}^R(x^S, t) = -d\, c_0^{-1}(\beta \times \alpha)\partial_t e^i(t)$$
$$\overset{*}{:} \int_{x' \in \partial \Omega^R} G(x'|x^S, t - \beta \cdot x'/c_0)\tau(x') \mathrm{d}l(x') \qquad (12.4)$$

where we have made use of (10.7) in (12.3) and

$$\mathcal{V}^B(x|x^S, t) = \mu_0 d\partial_t \mathcal{I}^B(t) * G(x|x^S, t) \qquad (12.5)$$

for the concentrated electric-current testing source (9.5) (see Eq. (2.11)). Here, $G(x|x^S, t)$ is the two-dimensional TD Green's function whose closed-form ray and modal representations for the rectangular domain Ω^R are given in Chapter 5. This strikingly straightforward relation makes it possible to evaluate the pulsed-voltage response to the incident impulsive plane. Owing to its simplicity it may be useful for estimating the (worst-case) pulsed radiated EM susceptibility of planar circuits.

It is interesting to note that Eq. (12.5) can also be derived from the contour-integral representation of the (time-dependent) far-field radiation characteristic (see Eq. (8.1))

$$\boldsymbol{E}_{\infty}^{B}(\boldsymbol{\xi}, t) = -d\, c_0^{-1} \mu_0 \partial_t^2 \mathcal{I}^B(t)$$

$$* \int_{\boldsymbol{x}' \in \partial\Omega^R} G^B(\boldsymbol{x}'|\boldsymbol{x}^S, t + \boldsymbol{\xi} \cdot \boldsymbol{x}'/c_0) \boldsymbol{\xi} \times \boldsymbol{\tau}(\boldsymbol{x}') \mathrm{d}l(\boldsymbol{x}') \quad (12.6)$$

and the property of self-reciprocity that relates the circuit's receiving and transmitting (or testing) states according to (10.14) and (10.15).

In the following sections, Eqs. (12.3) and (12.4) are applied to evaluate the plane-wave response $\mathcal{V}^R$ of an irregularly-shaped and a rectangular planar structure, respectively. For the former equation, the testing voltage distribution $\mathcal{V}^B$ is found via TD-CIM, while the latter makes use of the ray-type field representation, as given in Chapter 5.

12.2   Numerical results

The pulsed voltage response of both irregularly-shaped and rectangular planar structures is evaluated via Eqs. (12.3) and (12.4), respectively, and the referential FIT as implemented in CST Microwave Studio®. The circuits are irradiated by the (impulsive) incident plane wave that is defined via its polarization and propagation vectors (cf. Eqs. (10.6)–(10.7))

$$\boldsymbol{\alpha} = [\cos(\phi)\cos(\theta), \sin(\phi)\cos(\theta), -\sin(\theta)] \quad (12.7)$$

$$\boldsymbol{\beta} = [-\cos(\phi)\sin(\theta), -\sin(\phi)\sin(\theta), -\cos(\theta)] \quad (12.8)$$

respectively, and its (bipolar) triangular pulse shape (cf. (F.6))

$$e^i(t) = \frac{2e_{\mathrm{m}}}{t_{\mathrm{w}}} \left[t\,\mathrm{H}(t) - 2\left(t - \frac{t_{\mathrm{w}}}{2}\right)\mathrm{H}\left(t - \frac{t_{\mathrm{w}}}{2}\right) \right.$$

$$\left. + 2\left(t - \frac{3t_{\mathrm{w}}}{2}\right)\mathrm{H}\left(t - \frac{3t_{\mathrm{w}}}{2}\right) - (t - 2t_{\mathrm{w}})\,\mathrm{H}\left(t - 2t_{\mathrm{w}}\right) \right] \quad (12.9)$$

as shown in Fig. 12.1. Its amplitude is taken as $e_{\mathrm{m}} = 1 \cdot 10^3$ [V/m] and the zero-crossing time as $ct_{\mathrm{w}} = 1.0\, D/\sqrt{\epsilon_{\mathrm{r}}}$, where $c = c_0/\sqrt{\epsilon_{\mathrm{r}}}$. The reference FIT-based models are finely meshed, such that the maximum mesh step is always less than $ct_{\mathrm{w}}/50$. The boundary condition on the surrounding box is set to 'open'. Two different plane-wave excitations will be considered (a) $\{\phi, \theta\} = \{\pi/2, \pi/4\}$ (PW 1); (b) $\{\phi, \theta\} = \{\pi/4, \pi/4\}$

(PW 2). The resulting voltage responses are observed in the time window of observation $\{0 < ct \leq 10D/\sqrt{\epsilon_\mathrm{r}}\}$. In the following examples we take $\epsilon_\mathrm{r} = 4.50$, $D = 50.0\,[\mathrm{mm}]$ and $d = 1.50\,[\mathrm{mm}]$ circuit thickness.

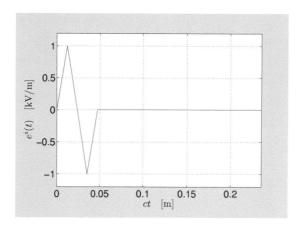

Figure 12.1: The triangular plane-wave signature. © [2016] Taylor & Francis Ltd. Adapted, with permission, from [130].

12.2.1 An irregularly-shaped planar circuit

At first, the pulsed response of the irregularly-shaped circuit, as shown in Fig. 12.2a, is evaluated via Eq. (12.3). It is assumed that at $t = 0$ plane wave **PW 1** hits the top edge $x_2 = 140\,[\mathrm{mm}]$, while **PW 2** at that origin hits the top-left corner $\{x_1, x_2\} = \{120, 140\}\,[\mathrm{mm}]$. The pulsed voltage responses are observed at $\{x_1^S, x_2^S\} = \{50, 100\}\,[\mathrm{mm}]$ (**PROBE 1**) and $\{x_1^S, x_2^S\} = \{30, 10\}\,[\mathrm{mm}]$ (**PROBE 2**) (see Fig. 12.2a).

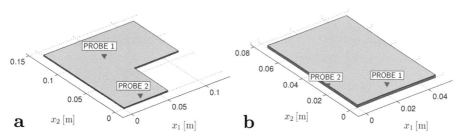

Figure 12.2: Computational models of the analysed circuits with field probes (the solid triangles). (a) The irregularly-shaped circuit; (b) The rectangular circuit.

The corresponding results are shown in Figs. (12.3) and (12.4). As can be seen, the pulses evaluated via Eq. (12.3) correlate very well with the FIT-based ones. The observable discrepancies may be dominantly attributed to different strategies in modeling the circuit's boundary conditions. While the TD-CIM-based model assumes the perfect magnetic wall along $\partial\Omega^R$, the three-dimensional FIT model has the 'open boundary' accounting for the fringing fields.

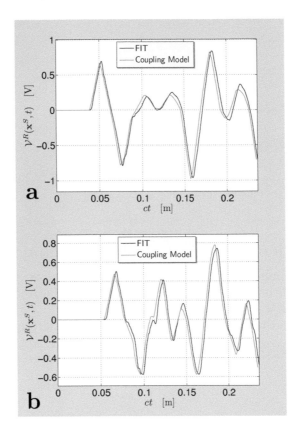

Figure 12.3: The pulsed voltage evaluated using the proposed coupling model and the referential FIT as observed at PROBE 1 of the irregularly-shaped circuit due to the plane wave (a) $\{\phi, \theta\} = \{\pi/2, \pi/4\}$ (PW 1); (b) $\{\phi, \theta\} = \{\pi/4, \pi/4\}$ (PW 2). © [2016] Taylor & Francis Ltd. Adapted, with permission, from [130].

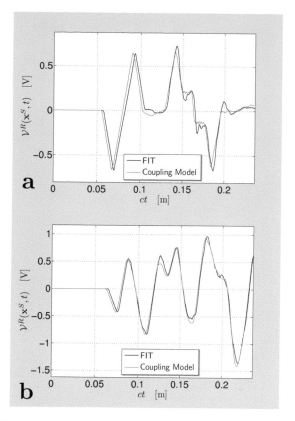

Figure 12.4: The pulsed voltage evaluated using the proposed coupling model and the referential FIT as observed at **PROBE** 2 of the irregularly-shaped circuit due to the plane wave (a) $\{\phi, \theta\} = \{\pi/2, \pi/4\}$ (PW 1); (b) $\{\phi, \theta\} = \{\pi/4, \pi/4\}$ (PW 2). © [2016] Taylor & Francis Ltd. Adapted, with permission, from [130].

12.2.2 A rectangular planar circuit

The pulsed voltage response of a rectangular circuit $\Omega = \{0 < x_1 < L, 0 < x_2 < W\}$ of dimensions $\{L, W\} = \{50, 75\}$ [mm] (see Fig. 12.2b) is evaluated with the help of (12.4) and (5.7) with (5.16). In this example, we assume conductive losses incorporated in the dielectric relaxation function according to Eq. (5.14) with $\sigma = 0.02$ [S/m]. The pulsed voltage responses are evaluated at two observation points $\{x_1^S, x_2^S\} = \{L/2, W/10\}$ (PROBE 1) and $\{x_1^S, x_2^S\} = \{0, W/3\}$ (PROBE 2) (see Fig. 12.2b.). It is assumed that at $t = 0$ plane wave PW 1 hits the top edge $x_2 = 75$ [mm], while PW 2 at that origin hits the top-left corner $\{x_1, x_2\} = \{50, 75\}$ [mm].

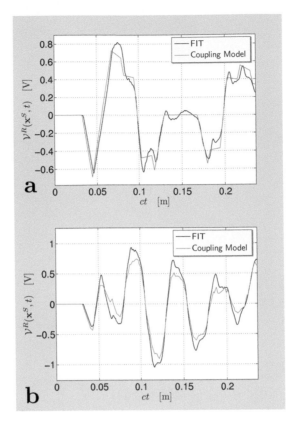

Figure 12.5: The pulsed voltage evaluated using the proposed coupling model and the referential FIT as observed at PROBE 1 of the rectangular circuit due to the plane wave (a) $\{\phi, \theta\} = \{\pi/2, \pi/4\}$ (PW 1); (b) $\{\phi, \theta\} = \{\pi/4, \pi/4\}$ (PW 2). © [2016] Taylor & Francis Ltd. Adapted, with permission, from [130].

The corresponding results are given in Figs. 12.5 and 12.6. Here, again, the differences, with respect to the FIT-based results, can be attributed mainly to the different boundary conditions imposed along the circuit periphery $\partial\Omega^R$. In this respect, it is interesting to interpret the discrepancies in the early part of the response in Fig. 12.6a. Owing to the fact that the incident magnetic-field vector is perpendicular to the tangential vector τ of the circuit's edges along $x_1 = 0$ and $x_1 = L$, these edges, where PROBE 2 is also located, do not contribute to the response (cf. Eq. (12.4)). On the other hand, with the FIT-based 'open-boundary model', the neighborhood of the observation probe also yields the (relatively weak) contribution with its (scaled) arrival time at about $cT_{\mathrm{arr}} = (2W/3)\sin(\pi/4)/\sqrt{\epsilon_{\mathrm{r}}} \simeq 0.0167$ [m]. The latter contribution ob-

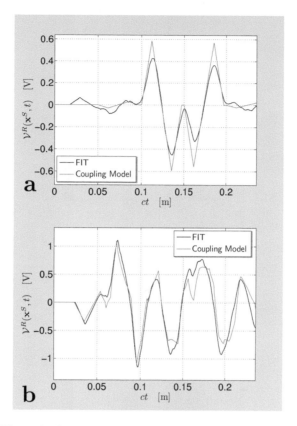

Figure 12.6: The pulsed voltage evaluated using the proposed coupling model and the referential FIT as observed at **PROBE 2** of the rectangular circuit due to the plane wave (a) $\{\phi, \theta\} = \{\pi/2, \pi/4\}$ (**PW 1**); (b) $\{\phi, \theta\} = \{\pi/4, \pi/4\}$ (**PW 2**). © [2016] Taylor & Francis Ltd. Adapted, with permission, from [130].

served at **PROBE 2** manifests itself earlier than the contributions from the edges along $x_2 = \{0, W\}$ that form the response of the analytical coupling model.

As the last example, the time evolution of the pulsed voltage distribution is illustrated. To this end, the TD response to the second plane wave, $\{\phi, \theta\} = \{\pi/4, \pi/4\}$ (**PW 2**), is evaluated for a set of field points on Ω^R. The results for the consecutive (scaled) observation times $ct = \{\sqrt{2}/30, \sqrt{2}/20, \sqrt{2}/15\}$ [m] are given in Fig. 12.7. Such analysis may help to localize 'hot spots' when assessing the vulnerability of PCBs to an external pulsed EM disturbance. Moreover, thanks to the fact that the pulsed voltage response has been expressed in the closed form whose evaluation is computationally very efficient, the result can

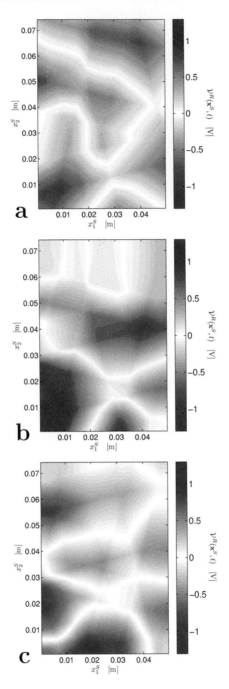

Figure 12.7: Time evolution of the voltage distribution at (a) $ct = \sqrt{2}/30\,[\mathrm{m}]$; (b) $ct = \sqrt{2}/20\,[\mathrm{m}]$; (c) $ct = \sqrt{2}/15\,[\mathrm{m}]$. © [2016] Taylor & Francis Ltd. Adapted, with permission, from [130].

find its application in solving optimization problems (e.g. [56]), which mostly necessitates (time-consuming) repeated calculations of the objective function.

12.3   Conclusions

Closed-form expressions for the efficient TD radiated susceptibility analysis of a planar circuit have been derived using the concept of TD reciprocity. From them it is immediately clear that a planar circuit can, in its receiving state, be viewed as being excited along the circuit periphery via the tangential component of an incident magnetic field.

The derived relations express the pulsed voltage response of a planar structure to an impulsive EM plane wave via the one-dimensional contour integration taken along the circuit's rim. It has been demonstrated that the presented approach is well suited for its combination with TD-CIM and the relevant TD 'ray-type' Green's function representation. In addition to the high computational efficiency of the presented methodology, the derived TD integral representations provide physical insights into the dominant (space-time) EM-coupling mechanism of planar structures. All the derived integral representations have been validated using (three-dimensional) FIT.

Chapter 13

Scattering Reciprocity Properties of an N-port Planar Circuit

The compensation theorem is a standard tool of linear-network theory that makes it possible to efficiently evaluate the impact of varying network elements on currents in the circuit at hand [105, Chap. 2, §12]. Owing to the possibility of extending the compensation theorem to its general EM (vectorial) form [66], it can be naturally expected that one can find a general expression for the change of N-port planar circuit's scattering properties due to the change in its loading elements. The confirmation of this expectation is the main result of this chapter, where such an expression is derived entirely in the TD with the aid of the EM reciprocity theorem of the time-convolution type (see [24, Sec. 28.2] and [133, Sec. 1.4.1]). More specifically, it is shown how the change in planar circuit's loading (lumped) elements influences such circuit's TD EM scattering characteristics in the receiving state of operation.

The following sections are organized as follows. At first, the relevant receiving situations are specified in Sec. 13.2. Then in Sec. 13.3, the transmitting and receiving scenarios are mutually interrelated in a way that yields the reciprocity relation describing the change of the scattered field due to the change in the circuit's load. Two special important cases pertaining to the N-port Thévenin and Norton equivalent circuits are briefly discussed. Finally, pulsed EM scattering properties of a multi-port planar circuit are analysed using the introduced TD compensation theorem. The obtained results are validated using FIT.

13.1 Model definition

Let us assume an N-port planar circuit that occupies a domain $\mathcal{D} \subset \mathbb{R}^3$ bounded by $\partial\mathcal{D} \subset \mathbb{R}^2$. The planar circuit under consideration is placed in the linear, homogeneous and isotropic embedding $\mathcal{D}_0$ whose EM properties are described by its electric permittivity ϵ_0 and magnetic permeability μ_0. The corresponding EM wave speed is $c_0 = (\epsilon_0\mu_0)^{-1/2} > 0$. The planar circuit is supposed to be causal and reciprocal in its EM behavior.

The planar circuit operates either in the transmitting state, in which it radiates the pulsed EM field into the embedding, or in the receiving state, in which is irradiated by a uniform pulsed-EM plane wave. In the receiving state, the circuit is supposed to be loaded by a number of discrete elements. The purpose of the following sections is to describe the impact of such varying load elements on circuit's pulsed-EM scattering behavior. For the corresponding transmitting situation we refer the reader to Sec. 11.2.

13.2 Receiving situations of an N-port planar circuit

The N-port planar circuit is in its receiving state activated by a uniform impulsive plane wave defined in Eqs. (10.6) and (10.7). In the analysis that follows we distinguish between two receiving scenarios, differing from each other in the circuit's load only. In the first one (see Fig. 13.1a), the circuit is at $\boldsymbol{x} = \boldsymbol{x}^n$ loaded by impedance $\mathcal{Z}^{L;n}$ for $n = \{1, \ldots, N\}$. The corresponding scattered field is defined as

$$\{\boldsymbol{E}^s, \boldsymbol{H}^s\} = \{\boldsymbol{E}^R - \boldsymbol{E}^i, \boldsymbol{H}^R - \boldsymbol{H}^i\} \tag{13.1}$$

and the voltage across the (linear) load is linearly related to the load current according to

$$\mathcal{V}^{R;n}(t) = \mathcal{Z}^{L;n}(t) * \mathcal{I}^{R;n}(t) \tag{13.2}$$

for $n = \{1, \ldots, N\}$ and $t > 0$. In the second case (see Fig. 13.1b), the circuit's load is changed to $\tilde{\mathcal{Z}}^{L;n}$. Note that this change does not

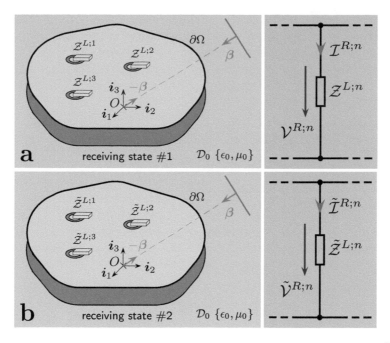

Figure 13.1: Receiving situations. The receiving circuit loaded by (a) $\mathcal{Z}^{L;n}$ for $n = \{1, 2, 3\}$; (b) $\tilde{\mathcal{Z}}^{L;n}$ for $n = \{1, 2, 3\}$.

have to be necessarily small. The corresponding scattered field is then defined as

$$\{\tilde{\boldsymbol{E}}^s, \tilde{\boldsymbol{H}}^s\} = \{\tilde{\boldsymbol{E}}^R - \boldsymbol{E}^i, \tilde{\boldsymbol{H}}^R - \boldsymbol{H}^i\} \tag{13.3}$$

The corresponding voltage across the load is then written as

$$\tilde{\mathcal{V}}^{R;n}(t) = \tilde{\mathcal{Z}}^{L;n}(t) * \tilde{\mathcal{I}}^{R;n}(t) \tag{13.4}$$

for $n = \{1, \ldots, N\}$ and $t > 0$. The objective of this chapter is to evaluate the change of the circuit's scattering properties, i.e.

$$\{\triangle \boldsymbol{E}^s, \triangle \boldsymbol{H}^s\} = \{\tilde{\boldsymbol{E}}^s, \tilde{\boldsymbol{H}}^s\} - \{\boldsymbol{E}^s, \boldsymbol{H}^s\} \tag{13.5}$$

with the aid of the TD reciprocity theorem. It will be shown that this change is intimately related to the circuit's transmitting state.

13.3 Reciprocity relations

In the first step, the scattered-field difference $\{\triangle \boldsymbol{E}^s, \triangle \boldsymbol{H}^s\}$ is interrelated with the (causal) testing wave field $\{\boldsymbol{E}^B, \boldsymbol{H}^B\}$ that is activated via the concentrated electric-current volume density $\boldsymbol{J}^B(\boldsymbol{x}|\boldsymbol{x}^B, t) = \mathcal{I}^B(t)\boldsymbol{\ell}\delta(\boldsymbol{x} - \boldsymbol{x}^B)$, $\boldsymbol{x}^B \in \mathcal{D}_0$, and whose constitutive EM properties are adjoint with respect to the ones pertaining to $\{\triangle\boldsymbol{E}^s, \triangle\boldsymbol{H}^s\}$. Consequently, application of the reciprocity theorem of the time-convolution type to the both states and to the unbounded domain $\mathcal{D}_0$ delivers the following relation

$$\int_{\boldsymbol{x}\in\partial\mathcal{D}} \left[\boldsymbol{E}^B(\boldsymbol{x}, t) \overset{*}{\times} \triangle\boldsymbol{H}^s(\boldsymbol{x}, t) - \triangle\boldsymbol{E}^s(\boldsymbol{x}, t) \overset{*}{\times} \boldsymbol{H}^B(\boldsymbol{x}, t) \right] \cdot \boldsymbol{\nu}(\boldsymbol{x}) \mathrm{d}A(\boldsymbol{x})$$

$$= -\int_{\boldsymbol{x}\in\mathcal{D}_0} \boldsymbol{J}^B(\boldsymbol{x}, t) \overset{*}{\cdot} \triangle\boldsymbol{E}^s(\boldsymbol{x}, t)\mathrm{d}V(\boldsymbol{x}) \qquad (13.6)$$

In virtue of the thin-slab approximation and thanks to the fact that the change of the scattered field is equal to the change of the total field, i.e. $\{\triangle\boldsymbol{E}^s, \triangle\boldsymbol{H}^s\} = \{\triangle\boldsymbol{E}^R, \triangle\boldsymbol{H}^R\}$, Eq. (13.6) can be rewritten as

$$\mathcal{I}^B(t) * \boldsymbol{\ell} \cdot \triangle\boldsymbol{E}^s(\boldsymbol{x}^B, t) = -\sum_{n=1}^{N} \mathcal{V}^{B;n}(t) * \triangle\mathcal{I}^{R;n}(t) \qquad (13.7)$$

for $\boldsymbol{x}^B \in \mathcal{D}_0$ and $t > 0$, where $\triangle\mathcal{I}^{R;n} = \tilde{\mathcal{I}}^{R;n} - \mathcal{I}^{R;n}$ (see Fig. 13.1). The remaining task is to relate the testing voltage $\mathcal{V}^{B;n}$ to the transmitting state. To find such a relation, the TD reciprocity theorem is now applied to the transmitting and testing states and to $\mathcal{D}_0$, which yields

$$\int_{\boldsymbol{x}\in\partial\mathcal{D}} \left[\boldsymbol{E}^T(\boldsymbol{x}, t) \overset{*}{\times} \boldsymbol{H}^B(\boldsymbol{x}, t) - \boldsymbol{E}^B(\boldsymbol{x}, t) \overset{*}{\times} \boldsymbol{H}^T(\boldsymbol{x}, t) \right] \cdot \boldsymbol{\nu}(\boldsymbol{x}) \mathrm{d}A(\boldsymbol{x})$$

$$= \int_{\boldsymbol{x}\in\mathcal{D}_0} \boldsymbol{J}^B(\boldsymbol{x}, t) \overset{*}{\cdot} \boldsymbol{E}^T(\boldsymbol{x}, t)\mathrm{d}V(\boldsymbol{x}) \qquad (13.8)$$

Finally it is noted that Eq. (13.8) can be simplified under the thin-slab approximation that is

$$\mathcal{I}^B(t) * \boldsymbol{\ell} \cdot \boldsymbol{E}^T(\boldsymbol{x}^B, t)|_{\mathcal{I}^{T;n}(t)=\delta(t)} * \mathcal{I}^{T;n}(t) = \mathcal{V}^{B;n}(t) * \mathcal{I}^{T;n}(t) \quad (13.9)$$

for $\boldsymbol{x}^B \in \mathcal{D}_0$ and $t > 0$, which upon combining with Eq. (13.7) results in

$$\triangle\boldsymbol{E}^s(\boldsymbol{x}^B, t) = -\sum_{n=1}^{N} \triangle\mathcal{I}^{R;n}(t) * \boldsymbol{E}^T(\boldsymbol{x}^B, t)|_{\mathcal{I}^{T;n}(t)=\delta(t)} \qquad (13.10)$$

for $\boldsymbol{x}^B \in \mathcal{D}_0$ and $t > 0$, where we have invoked the condition that the resulting relation holds true for arbitrary $\mathcal{I}^B$. Owing to the fact that the both scattered and radiated fields admit the far-field expansion for (causal) outgoing wave fields (see Eq. (11.2)), Eq. (13.10) can be further rewritten using the far-field amplitudes, that is

$$\triangle \boldsymbol{E}_\infty^s(\boldsymbol{\xi}, t) = -\sum_{n=1}^{N} \triangle \mathcal{I}^{R;n}(t) * \boldsymbol{E}_\infty^T(\boldsymbol{\xi}, t)|_{\mathcal{I}^{T;n}(t) = \delta(t)} \qquad (13.11)$$

for $\boldsymbol{\xi} \in \Omega = \{\boldsymbol{\xi} \cdot \boldsymbol{\xi} = 1\}$ and $t > 0$, where $\boldsymbol{\xi} = \boldsymbol{x}/|\boldsymbol{x}|$ denotes the unit vector in the direction of observation.

Finally note that in line with the corresponding real-FD circuit counterpart [51], two special forms of the TD compensation theorem (13.10) can be put forward. Namely, if the reference receiving circuit from Fig. 13.1a is open-circuited at its ports, we get

$$\triangle \boldsymbol{E}^s(\boldsymbol{x}^B, t) = -\sum_{n=1}^{N} \tilde{\mathcal{I}}^{R;n}(t) * \boldsymbol{E}^T(\boldsymbol{x}^B, t)|_{\mathcal{I}^{T;n}(t) = \delta(t)} \qquad (13.12)$$

for $\boldsymbol{x}^B \in \mathcal{D}_0$ and $t > 0$, while the short-circuited reference leads to

$$\triangle \boldsymbol{E}^s(\boldsymbol{x}^B, t) = -\sum_{n=1}^{N} [\tilde{\mathcal{I}}^{R;n}(t) - \mathcal{I}^{G;n}(t)] * \boldsymbol{E}^T(\boldsymbol{x}^B, t)|_{\mathcal{I}^{T;n}(t) = \delta(t)} \quad (13.13)$$

for $\boldsymbol{x}^B \in \mathcal{D}^\infty$ and $t > 0$, where $\tilde{I}^{G;n}$ is Norton's short-circuit current at the n-th port. The real-FD counterparts of special cases (13.12)–(13.13) for $N = 1$ have been previously applied to discussing the adequacy of Kirchhoff's network representation of a receiving antenna [16, Eqs. (9) and (10)]. The corresponding reciprocity analysis concerning an N-port antenna system has been carried out in [133, Ch. 9] at a somewhat more general level.

13.4 Numerical results

In order to validate the introduced TD compensation theorem, pulsed EM scattering from a 3-port planar circuit is analysed numerically. Specifically, the analysed structure is shown in Fig. 13.2. Its thickness is $d = 1.50\,[\text{mm}]$ and the relative permittivity of the dielectric slab is $\epsilon_r = 4.0$, which implies that the corresponding EM wave speed is $c = c_0/2$. The circuit is loaded by discrete lumped elements at its accessible ports. In the analysis that follows, we shall use resistive $\mathcal{Z}^{L;n} = R^{L;n}\delta(t)$ and capacitive $\mathcal{Z}^{L;n} = (C^{L;n})^{-1}\text{H}(t)$ load lumped elements. The definition of the circuit's loads in both receiving scenarios is summarized in Table 13.1.

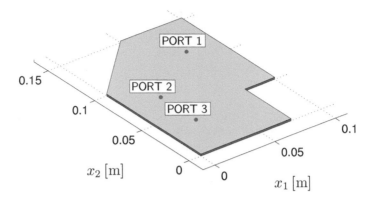

Figure 13.2: Computational model of the analysed circuit with three accessible ports (the dots on the patch). © [2018] IEEE. Adapted, with permission, from [134].

Table 13.1: Definition of the circuit's load.

port	position [mm]	$\mathcal{Z}^{L;n}$ [Ω/s]	$\tilde{\mathcal{Z}}^{L;n}$ [Ω/s]
$n = 1$	$[75.0, 112.5, 0]$	$10^{-6}\delta(t)$	$10^2\delta(t)$
$n = 2$	$[25.0, 75.0, 0]$	$5 \cdot 10^{10}\mathrm{H}(t)$	$5 \cdot 10^2\delta(t)$
$n = 3$	$[25.0, 37.5, 0]$	$10^2\delta(t)$	$2 \cdot 10^2\delta(t)$

The circuit is, in the receiving scenario, irradiated by the incident plane wave (see Eqs. (10.6) and (10.7)) defined via

$$\boldsymbol{\alpha} = [0, \cos(\gamma), \sin(\gamma)] \tag{13.14}$$
$$\boldsymbol{\beta} = [0, -\sin(\gamma), \cos(\gamma)] \tag{13.15}$$

where we take $\gamma = 3\pi/4$, for example, and its (bipolar) triangular signature (see Eq. (12.9)) with $e_\mathrm{m} = 1.0 \cdot 10^3$ [V/m] and $ct_\mathrm{w} = 0.25$ [m] (see Fig. 13.3). The main goal of the experiment is to evaluate the change of the co-polarized back-scattered-field amplitude in the far-field region according to Eq. (13.11). To this end, the circuit is first analysed in its transmitting state using TD-CIM that readily provides the (time-integrated) pulsed radiation amplitude (see Sec. 8.2). In the transmitting situation, the circuit is excited at its accessible ports by the electric-

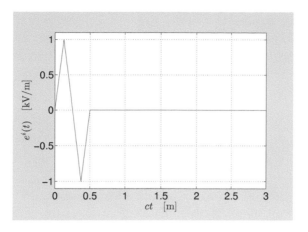

Figure 13.3: The triangular plane-wave signature.

current pulse of the (unipolar) triangular pulse shape (cf. Eq. (5.24))

$$\mathcal{I}^{T;n}(t) = 2A\left[\frac{t}{t_{\rm w}}{\rm H}(t) - 2\left(\frac{t}{t_{\rm w}} - \frac{1}{2}\right){\rm H}\left(\frac{t}{t_{\rm w}} - \frac{1}{2}\right)\right.$$

$$\left. + \left(\frac{t}{t_{\rm w}} - 1\right){\rm H}\left(\frac{t}{t_{\rm w}} - 1\right)\right] \tag{13.16}$$

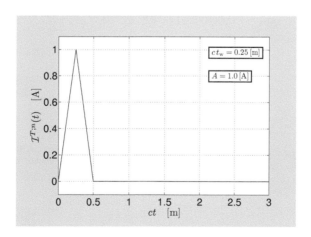

Figure 13.4: The triangular excitation signature.

for $n = \{1, 2, 3\}$, with $A = 1.0\,[{\rm A}]$ and $ct_{\rm w} = 0.50\,[{\rm m}]$ (see Fig. 13.4). The resulting radiated pulse shapes in the backward direction, that is at $\boldsymbol{\xi} = -\boldsymbol{\beta}$, are shown in Fig. 13.5. The radiation amplitudes in Eq. (13.11) call for a deconvolution procedure, which is a tough problem in general. For the chosen triangular pulse, however, an efficient and stable algorithm does exist. The relevant procedure can be described as

$$\mathbf{I}_t^2 \triangle \boldsymbol{E}_\infty^T(-\boldsymbol{\beta}, t)|_{\mathcal{I}^{T;n}(t)=\delta(t)} = (t_{\mathrm{w}}/2A) \sum_{n=0}^{\infty} (n+1)$$

$$\partial_t \left[\mathbf{I}_t \boldsymbol{E}_\infty^T(-\boldsymbol{\beta}, t - nt_{\mathrm{w}}/2)|_{\mathcal{I}^{T;n}(t)=\delta(t)} * \mathcal{I}^{T;n}(t) \right] \tag{13.17}$$

Finally, the desired radiation amplitudes are obtained upon differentiating the result. The number of terms in the sum of Eq. (13.17) is in any bounded time window of observation finite and its evaluation does not introduce any difficulties.

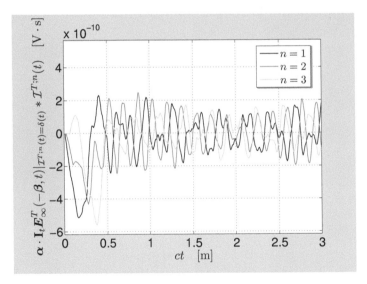

Figure 13.5: The time-integrated, co-polarized radiated pulse shapes in the backward direction. © [2018] IEEE. Adapted, with permission, from [134].

The receiving scenarios (see Table 13.1) have been analysed using the FIT as implemented in CST Microwave Studio®. The calculated electric-current differences $\triangle I^{R;n}$ are given in Fig. 13.6. As can be seen, the smallest electric-current difference is observed at **PORT 3**, where we have changed its load resistance from $100\,\Omega$ to $200\,\Omega$.

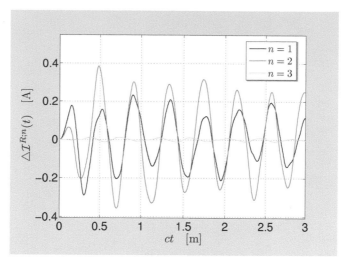

Figure 13.6: Load electric-current differences. © [2018] IEEE. Adapted, with permission, from [134].

With the impulse radiation characteristics and the electric-current difference at our disposal, the change in the TD back-scattering amplitude can be found upon evaluating the sum of their time convolutions Eq. (13.11). This can be accomplished via the standard trapezoidal rule, for instance. In this way, we can evaluate the change of the back-scattered field amplitudes separately for each change in the circuit's load. The corresponding pulse contributions are given in Fig. 13.7a. As expected, the lowest impact on the circuit-scattering variation has the load change at **PORT 3** (cf. Fig. 13.6 and Table 13.1).

Finally, in order to validate the results, the sum of the contributions from Fig. 13.7a has been compared with the difference of the actual back-scattering amplitudes calculated using the 'far-field probes' of CST Microwave Studio®. Fig. 13.7b shows the resulting pulses. Given the delicacy of the scattered-field difference, the resulting pulse shapes correlate well, thus validating the introduced TD compensation theorem. The visible discrepancies can largely be attributed to the 'signal processing' in Eqs. (13.11) and (13.17) as well as to unavoidable errors introduced by the used numerical techniques. In any case, the main goal of the numerical experiment, that is, the validation of the main result (13.11), has been achieved.

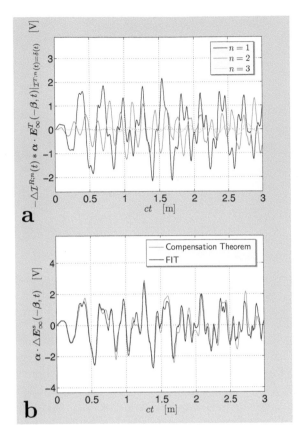

Figure 13.7: The co-polarized back-scattered field difference. (a) Contributions from the particular load changes; (b) the total field as evaluated using the proposed compensation theorem and the referential FIT. © [2018] IEEE. Adapted, with permission, from [134].

13.5 Conclusions

The impact of the change of lumped loads on TD N-port planar-circuit's scattering properties has been studied with the aid of the reciprocity theorem of the time-convolution type. The result explicitly shows how this change in the circuit's EM scattering is related to (the sum of) the TD radiated EM fields of the corresponding transmitting situations. Besides gaining physical insights into the scattering of N-port planar structures, the derived results provide the theoretical basis for controlling the pulsed echo of multi-port planar scatterers and for TD scattering antenna measurement techniques. By virtue of the applied methodology based on the TD reciprocity theorem, the compensation theorem can be applied to the planar circuits with varying loads, which makes it suitable for analyzing the operation of passive radio-frequency identification (RFID) tags whose pulsed EM backscattering is modulated by switching between their (chip's) loads (see e.g. [77]). For more numerical examples we refer the reader to [134].

Chapter 14

Scattering of Conductive and Dielectric Inclusions

Owing to the unprecedented expansion of digital communication devices, there is a need for fundamentally new modeling techniques that would account for their pulsed (bit-like) mode of operation with accompanying space-time EM interference phenomena. As for the compensation theorems and their applications, they have been limited to real-FD so far (e.g. [66]). Accordingly, this chapter introduces TD compensation theorems concerning planar circuits and provides efficient means for such TD EM modeling. More specifically, the Lorentz reciprocity theorem of the time-convolution type (see [24, Sec. 28.2] and [133, Sec. 1.4.1]) is applied to describe the impact of an inclusion on the pulsed EM-field response of a planar circuit. As such, the introduced compensation theorems provide a convenient starting point for TD analysis of inclusions in planar structures, such as the photonic crystal power/ground layer [142] or substrate-integrated waveguides [95]. In contrast to what one can gain from brute-force numerical modeling, the theorems may lead to numerous variational forms and approximate expressions, revealing physical insights that are necessary for efficient engineering practice.

The ensuing sections are organized as follows. After formulating the problem in Sec. 14.1, the generic reciprocity relation is derived in Sec. 14.2. Subsequently, the TD compensation theorems for conductive and dielectric inclusions in planar circuits are constructed in Sec. 14.3. Application of the derived TD theorems is the subject of the following Sec. 14.4. Finally, in order to demonstrate the applicability of the introduced theory, the impact of a shorting circular pin in a planar circuit is analysed in Sec. 14.5. Here, the obtained TD results are validated using FIT.

14.1   Problem definition

The planar circuit under consideration is placed in the linear, homogeneous and isotropic embedding $\mathcal{D}_0$ whose EM properties are described by its electric permittivity ϵ_0 and magnetic permeability μ_0. The corresponding EM wave speed is $c_0 = (\epsilon_0\mu_0)^{-1/2} > 0$. Its thickness d is assumed to be small with respect to the spatial support of the excitation pulse, such that the cavity model applies (see Sec. 2.1). The EM properties of the circuit are described by its dielectric relaxation function $\kappa(t)$ and (instantaneously-reacting) magnetic relaxation function $\mu_0\delta(t)$. The perfectly conducting plates of the circuit occupy a surface domain Ω that is closed by the boundary contour $\partial\Omega$ with ν being its outer normal vector. It is assumed that within the circuit there is an inclusion of bounded volume with dielectric and/or conductive properties whose EM constitutive properties differ from its surrounding. The dielectric and conduction relaxation functions of the inclusion are $\kappa^P(\boldsymbol{x}, t)$ and $\sigma^P(\boldsymbol{x}, t)$, respectively. The surface domain pertaining to the inclusion is denoted by Γ and its bounding contour is $\partial\Gamma$. The relevant normal vector ν is oriented outwards (see Fig. 14.1).

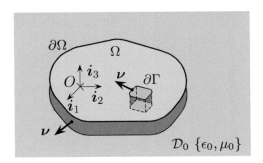

Figure 14.1: Planar circuit with an inclusion.

In the analysis that follows, we distinguish between the field quantities $\{E_3^A, \boldsymbol{H}^A\}$ that would be excited in absence of the inclusion (state A) and the actual field quantities $\{\tilde{E}_3^A, \tilde{\boldsymbol{H}}^A\}$ that account for its presence (state $\tilde{A}$). Both field states are excited by the electric-current surface density J_3^A with $\mathrm{supp}(J_3^A) \subset \Omega \setminus \Gamma$, i.e. exterior to the domain of the inclusion. The incremental change (not necessarily infinitesimally small) of the actual field quantities is defined as $\{E_3^\Delta, \boldsymbol{H}^\Delta\} = \{\tilde{E}_3^A - E_3^A, \tilde{\boldsymbol{H}}^A - \boldsymbol{H}^A\}$ and the corresponding field state is denoted by ΔA. The actual field states are interrelated with the testing field state $\{E_3^B, \boldsymbol{H}^B\}$ (state B) that is activated by the concentrated source $J_3^B(\boldsymbol{x}|\boldsymbol{x}^S, t) = \mathcal{I}^B(t)\delta(\boldsymbol{x} - \boldsymbol{x}^S)$ at $\boldsymbol{x}^S \in \Omega$ and shows, in its constitutive parameters, no contrast with respect to state A. The actual, as well as the testing fields, do satisfy the magnetic-wall boundary condition along $\partial\Omega$. The electric-current surface density and the voltage are denoted by $\partial \boldsymbol{J} = -\boldsymbol{i}_3 \times \boldsymbol{H}$ and $\mathcal{V} = -dE_3$, respectively, with the relevant superscript.

14.2   Generic reciprocity relation

Construction of the TD compensation theorems introduced in this chapter is largely based on the generic reciprocity relation that is derived next. To this end we consider two EM-field states, say A and B, that satisfy (cf. Eqs. (2.1)–(2.3))

$$-\partial_1 H_2^{A,B} + \partial_2 H_1^{A,B} + \partial_t \kappa(t) * E_3^{A,B} = -J_3^{A,B} \tag{14.1}$$

$$\partial_2 E_3^{A,B} + \mu_0 \partial_t H_1^{A,B} = 0 \tag{14.2}$$

$$-\partial_1 E_3^{A,B} + \mu_0 \partial_t H_2^{A,B} = 0 \tag{14.3}$$

for all $\boldsymbol{x} \in \Omega \subset \mathbb{R}^2$ and all $t > 0$, where $\kappa(t)$ is the (causal) dielectric relaxation function. Upon integrating the relevant (local) interaction quantity over $\Lambda \subset \Omega$ and applying Gauss' divergence theorem we finally end up with (cf. Eq. (2.29))

$$\int_{\boldsymbol{x}' \in \partial\Lambda} \mathcal{V}^B(\boldsymbol{x}'|\boldsymbol{x}^S, t) * \boldsymbol{\nu}(\boldsymbol{x}') \cdot \partial \boldsymbol{J}^A(\boldsymbol{x}', t)\mathrm{d}l(\boldsymbol{x}')$$

$$-\int_{\boldsymbol{x}' \in \partial\Lambda} \mathcal{V}^A(\boldsymbol{x}', t) * \boldsymbol{\nu}(\boldsymbol{x}') \cdot \partial \boldsymbol{J}^B(\boldsymbol{x}'|\boldsymbol{x}^S, t)\mathrm{d}l(\boldsymbol{x}')$$

$$= \int_{\boldsymbol{x}' \in \Lambda} \mathcal{V}^B(\boldsymbol{x}'|\boldsymbol{x}^S, t) * J_3^A(\boldsymbol{x}', t)\mathrm{d}A(\boldsymbol{x}')$$

$$-\chi_\Lambda(\boldsymbol{x}^S)\int_{\boldsymbol{x}' \in \Lambda} \mathcal{V}^A(\boldsymbol{x}', t) * J_3^B(\boldsymbol{x}'|\boldsymbol{x}^S, t)\mathrm{d}A(\boldsymbol{x}') \tag{14.4}$$

where $\mathcal{V}^{A,B} = -dE_3^{A,B}$, $\partial \boldsymbol{J}^{A,B} = -\boldsymbol{i}_3 \times \boldsymbol{H}^{A,B}$, Λ is the generic domain with (sufficiently regular) boundary $\partial\Lambda$ and with the outward-oriented unit vector $\boldsymbol{\nu}$, $\chi_\Lambda(\boldsymbol{x})$ is the characteristic function of Λ and $\boldsymbol{x}^S$ specifies the position of the (testing) source. The global interaction quantity of the time-convolution type (14.4) is used in the analysis that follows.

14.3   TD compensation theorems

In this section, the TD compensation theorems are derived. To this end, the generic reciprocity relation of the time-convolution type (14.4) is systematically employed.

First, the generic reciprocity theorem is applied to the bounded domain $\Omega \setminus \Gamma$ and to ΔA and B states. Owing to the magnetic-wall boundary conditions, that is

$$\lim_{\delta \downarrow 0} \boldsymbol{\nu}(\boldsymbol{x}) \cdot \partial \boldsymbol{J}^{\Delta}(\boldsymbol{x} + \delta \boldsymbol{\nu}, t) = 0 \tag{14.5}$$

$$\lim_{\delta \downarrow 0} \boldsymbol{\nu}(\boldsymbol{x}) \cdot \partial \boldsymbol{J}^{B}(\boldsymbol{x} + \delta \boldsymbol{\nu} | \boldsymbol{x}^{S}, t) = 0 \tag{14.6}$$

for all $\boldsymbol{x} \in \partial \Omega$ and all $t > 0$, and due to the fact that the difference of the actual field states is source-free in $\Omega \setminus \Gamma$, we arrive at

$$\mathcal{I}^{B}(t) * \mathcal{V}^{\Delta}(\boldsymbol{x}^{S}, t) \chi_{\Omega \setminus \Gamma}(\boldsymbol{x}^{S})$$
$$= \int_{\boldsymbol{x}' \in \partial \Gamma} \mathcal{V}^{B}(\boldsymbol{x}' | \boldsymbol{x}^{S}, t) * \boldsymbol{\nu}(\boldsymbol{x}') \cdot \partial \boldsymbol{J}^{\Delta}(\boldsymbol{x}', t) \mathrm{d}l(\boldsymbol{x}')$$
$$- \int_{\boldsymbol{x}' \in \partial \Gamma} \mathcal{V}^{\Delta}(\boldsymbol{x}', t) * \boldsymbol{\nu}(\boldsymbol{x}') \cdot \partial \boldsymbol{J}^{B}(\boldsymbol{x}' | \boldsymbol{x}^{S}, t) \mathrm{d}l(\boldsymbol{x}') \tag{14.7}$$

for all $\boldsymbol{x}^{S} \in \Omega$ and all $t > 0$. Secondly, the reciprocity theorem is applied to the surface domain Γ occupied by the inclusion and to A and B states. Since the actual field state is source-free in Γ, we get the following relation

$$\int_{\boldsymbol{x}' \in \partial \Gamma} \mathcal{V}^{B}(\boldsymbol{x}' | \boldsymbol{x}^{S}, t) * \boldsymbol{\nu}(\boldsymbol{x}') \cdot \partial \boldsymbol{J}^{A}(\boldsymbol{x}', t) \mathrm{d}l(\boldsymbol{x}')$$
$$- \int_{\boldsymbol{x}' \in \partial \Gamma} \mathcal{V}^{A}(\boldsymbol{x}', t) * \boldsymbol{\nu}(\boldsymbol{x}') \cdot \partial \boldsymbol{J}^{B}(\boldsymbol{x}' | \boldsymbol{x}^{S}, t) \mathrm{d}l(\boldsymbol{x}')$$
$$= \mathcal{I}^{B}(t) * \mathcal{V}^{A}(\boldsymbol{x}^{S}, t) \chi_{\Gamma}(\boldsymbol{x}^{S}) \tag{14.8}$$

for all $\boldsymbol{x}^S \in \Omega$ and all $t > 0$. Making use of Eq. (14.8) in (14.7) yields *the first compensation theorem*, viz

$$\mathcal{I}^B(t) * \left[\mathcal{V}^\Delta(\boldsymbol{x}^S, t) \chi_{\Omega \backslash \Gamma}(\boldsymbol{x}^S) - \mathcal{V}^A(\boldsymbol{x}^S, t) \chi_\Gamma(\boldsymbol{x}^S) \right]$$

$$= \int_{\boldsymbol{x}' \in \partial\Gamma} \mathcal{V}^B(\boldsymbol{x}' | \boldsymbol{x}^S, t) * \boldsymbol{\nu}(\boldsymbol{x}') \cdot \partial \tilde{\boldsymbol{J}}^A(\boldsymbol{x}', t) \mathrm{d}l(\boldsymbol{x}')$$

$$- \int_{\boldsymbol{x}' \in \partial\Gamma} \tilde{\mathcal{V}}^A(\boldsymbol{x}', t) * \boldsymbol{\nu}(\boldsymbol{x}') \cdot \partial \boldsymbol{J}^B(\boldsymbol{x}' | \boldsymbol{x}^S, t) \mathrm{d}l(\boldsymbol{x}')$$

$$\tag{14.9}$$

for all $\boldsymbol{x}^S \in \Omega$ and all $t > 0$. Equation (14.8) describes the impact of the inclusion on the pulsed voltage response in terms of equivalent (unknown) current surface densities along the inclusion's boundary $\partial\Gamma$. The form of (14.9) is particularly suitable for handling electromagnetically impenetrable inclusions. For the latter, the explicit-type boundary conditions apply, viz

$$\lim_{\delta \downarrow 0} \tilde{\mathcal{V}}^A(\boldsymbol{x} + \delta\boldsymbol{\nu}, t) = 0 \quad \text{for} \quad \text{PEC} \tag{14.10}$$

$$\lim_{\delta \downarrow 0} \boldsymbol{\nu}(\boldsymbol{x}) \cdot \tilde{\boldsymbol{J}}^A(\boldsymbol{x} + \delta\boldsymbol{\nu}, t) = 0 \quad \text{for} \quad \text{PMC} \tag{14.11}$$

for all $\boldsymbol{x} \in \partial\Gamma$ and all $t > 0$.

In case of a penetrable inclusion, the reciprocity theorem is applied to the surface domain pertaining to the inclusion Γ and to ΔA and B states, which yields

$$\int_{\boldsymbol{x}' \in \partial\Gamma} \mathcal{V}^B(\boldsymbol{x}' | \boldsymbol{x}^S, t) * \boldsymbol{\nu}(\boldsymbol{x}') \cdot \partial \boldsymbol{J}^\Delta(\boldsymbol{x}', t) \mathrm{d}l(\boldsymbol{x}')$$

$$- \int_{\boldsymbol{x}' \in \partial\Gamma} \mathcal{V}^\Delta(\boldsymbol{x}', t) * \boldsymbol{\nu}(\boldsymbol{x}') \cdot \partial \boldsymbol{J}^B(\boldsymbol{x}' | \boldsymbol{x}^S, t) \mathrm{d}l(\boldsymbol{x}')$$

$$= \int_{\boldsymbol{x}' \in \Gamma} \mathcal{V}^B(\boldsymbol{x}' | \boldsymbol{x}^S, t) * J_3^C(\boldsymbol{x}', t) \mathrm{d}A(\boldsymbol{x}')$$

$$- \mathcal{I}^B(t) * \mathcal{V}^\Delta(\boldsymbol{x}^S, t) \chi_\Gamma(\boldsymbol{x}^S) \tag{14.12}$$

for all $\boldsymbol{x}^S \in \Omega$ and all $t > 0$, where the equivalent contrast-source electric-current volume density reads

$$J_3^C(\boldsymbol{x}, t) = \sigma^P(\boldsymbol{x}, t) * \tilde{E}_3^A(\boldsymbol{x}, t) + \partial_t [\kappa^P(\boldsymbol{x}, t) - \kappa(t)] * \tilde{E}_3^A(\boldsymbol{x}, t) \tag{14.13}$$

for all $\boldsymbol{x} \in \Gamma$ and all $t > 0$. Finally, combination of Eq. (14.7) with (14.12) leads to *the second compensation theorem*

$$\mathcal{I}^B(t) * \mathcal{V}^\Delta(\boldsymbol{x}^S, t) = \int_{\boldsymbol{x}' \in \Gamma} \mathcal{V}^B(\boldsymbol{x}' | \boldsymbol{x}^S, t) * J_3^C(\boldsymbol{x}', t) \mathrm{d}A(\boldsymbol{x}') \tag{14.14}$$

for all $\boldsymbol{x}^S \in \Omega$ and all $t > 0$. Equation (14.14) describes the impact of the electromagnetically penetrable inclusion on the pulsed voltage response in terms of the equivalent (unknown) contrast-source electric-current volume density. An application of the compensation theorem to TD analysis of 'perturbed' planar circuits is briefly described in the following section.

14.4 Application of the TD compensation theorems

The evaluation of the pulsed voltage change expressed using (14.9) and (14.14) is carried out in two steps. The first one requires solving a space-time integral equation while the second is a mere evaluation of the voltage difference. The procedure will be described next for PEC and penetrable inclusions.

14.4.1 The impact of a PEC inclusion

In the first step, the point of observation $\boldsymbol{x}^S$ in Eq. (14.9) is chosen such that it approaches $\partial\Gamma$ along which the boundary condition (14.10) applies. In this way we get

$$-\mathcal{I}^B(t) * \mathcal{V}^A(\boldsymbol{x}^S, t)$$
$$= \int_{\boldsymbol{x}' \in \partial\Gamma} \mathcal{V}^B(\boldsymbol{x}'|\boldsymbol{x}^S, t) * \boldsymbol{\nu}(\boldsymbol{x}') \cdot \partial \tilde{\boldsymbol{J}}^A(\boldsymbol{x}', t) \mathrm{d}l(\boldsymbol{x}') \quad (14.15)$$

for $\boldsymbol{x}^S \in \partial\Gamma$ and $t > 0$. Due to the fact that the voltage response $\mathcal{V}^A(\boldsymbol{x}^S, t)$ as well as the testing voltage distribution $\mathcal{V}^B(\boldsymbol{x}, t)$ can be readily evaluated for $\boldsymbol{x} \in \partial\Gamma$ and $t > 0$ using TD-CIM, Eq. (14.15) can be interpreted as a time-convolution, (spatially) Fredholm integral-equation of the first kind, in which (the normal component of) the space-time distribution of the electric-current surface density along $\partial\Gamma$ is unknown. Once the integral equation (14.15) is solved, Eq. (14.9) with (14.10) is applied again, but now to evaluate the voltage difference exterior to the inclusion, that is

$$\mathcal{I}^B(t) * \mathcal{V}^\Delta(\boldsymbol{x}^S, t) = \int_{\boldsymbol{x}' \in \partial\Gamma} \mathcal{V}^B(\boldsymbol{x}'|\boldsymbol{x}^S, t) * \boldsymbol{\nu}(\boldsymbol{x}') \cdot \partial \tilde{\boldsymbol{J}}^A(\boldsymbol{x}', t) \mathrm{d}l(\boldsymbol{x}')$$
$$(14.16)$$

for $\boldsymbol{x}^S \in \Omega \setminus \Gamma$ and $t > 0$. Since the voltage $\mathcal{V}^A(\boldsymbol{x}, t)$ in $\boldsymbol{x} \in \Omega \setminus \Gamma$ and $t > 0$ is easily attainable using TD-CIM, the pulsed voltage of the circuit with the PEC inclusion $\tilde{\mathcal{V}}^A(\boldsymbol{x}, t)$ immediately follows.

14.4.2 The impact of a penetrable inclusion

In order to evaluate the effect of a penetrable inclusion with conductive or/and dielectric properties, the point of observation x^S in Eq. (14.14) is first moved to Γ, which gives

$$\mathcal{I}^B(t) * \mathcal{V}^A(x^S, t) = \mathcal{I}^B(t) * \tilde{\mathcal{V}}^A(x^S, t)$$
$$- \int_{x' \in \Gamma} \mathcal{V}^B(x'|x^S, t) * J_3^C(x', t) \mathrm{d}A(x') \qquad (14.17)$$

for $x \in \Gamma$ and $t > 0$. Since the equivalent contrast-source electric-current volume density is a function of $\tilde{\mathcal{V}}^A$ (see Eq. (14.13)), Eq. (14.17) can be classified as a time-convolution, (spatially) Fredholm integral-equation of the second kind. Once the latter is solved for $J_3^C(x, t)$ with $x \in \Gamma$ and $t > 0$, Eq. (14.14) is applied again in order to evaluate the change in the pulsed voltage response for all $x^S \in \Omega$ and $t > 0$. Again, as the space-time distribution of $\mathcal{V}^A(x, t)$ in $x \in \Omega$ and $t > 0$ is readily available via TD-CIM, the pulsed voltage of the circuit with the penetrable inclusion $\tilde{\mathcal{V}}^A(x, t)$ follows.

Finally, note that if the testing source is placed at the position where the actual field is excited in the second step, one may evaluate the change of the input impedance due to an inclusion. A detailed description of implementation aspects with accompanying numerical results concerning a PEC inclusion is given in the following section.

14.5 Numerical results

The planar circuit in question is shown in Fig. 14.2. The thickness of the circuit is $d = 1.50\,[\mathrm{mm}]$. Its EM properties are taken to be instantaneously reacting with $\kappa(t) = 4.20\epsilon_0\,\delta(t)$. The corresponding EM wave speed is $c = c_0/\sqrt{4.20}$, where c_0 is the EM wave speed in a vacuum. The circuit is at $x^P = \{50.0, 112.5\}\,[\mathrm{mm}]$ excited by the vertical electric-current port (viz **PORT** in Fig. 14.2) with the (unipolar) triangular pulse shape defined in Eq. (5.24), where we take $A = 1.0\,[\mathrm{A}]$ and $ct_\mathrm{w} = 0.10\,[\mathrm{m}]$ (see Fig. 14.3). The pulsed-voltage response is observed at $x^S = \{75.0, 75.0\}\,[\mathrm{mm}]$ (see **PROBE** in Fig. 14.2) and within the finite window of observation $\{0 \le ct \le 2.0\}\,[\mathrm{m}]$. The main objective of this section is to evaluate the effect of a circular PEC pin of radius $\varrho = 1.0\,[\mathrm{mm}]$ whose center is at $x^C = \{25.0, 36.5\}\,[\mathrm{mm}]$ (see **PIN** in Fig. 14.2). To meet this goal, the procedure described in Sec. 14.4.1 is followed.

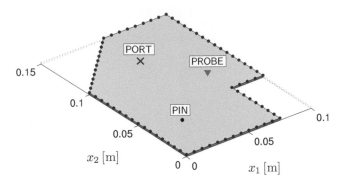

Figure 14.2: Computational model of the analysed circuit with an excitation port (the cross), a field probe (the solid triangle) and a PEC pin (the dot).

In the first step, we make use of Eq. (14.15) in order to find the electric-current surface density on the pin's circular contour $\partial\Gamma$. Owing to the fact that $\varrho \ll ct_{\mathrm{w}}$ we may assume the constant electric-current surface-density distribution $\partial J(t)$ along $\partial\Gamma$ and write down the following approximation

$$-\mathcal{I}^B(t) * \mathcal{V}^A(\boldsymbol{x}^S, t) \simeq 2\pi\varrho\, \partial J(t) * \mathcal{V}^B(\boldsymbol{x}'|\boldsymbol{x}^S, t) \qquad (14.18)$$

where $\boldsymbol{x}^S = \boldsymbol{x}^C + \{0, \varrho\}$ and $\boldsymbol{x}' = \boldsymbol{x}^C$. The testing voltage distribution is evaluated via (cf. Eq. (2.37))

$$\mathcal{V}^B(\boldsymbol{x}'|\boldsymbol{x}^S, t) = \mu_0 d\partial_t\, \mathcal{I}^B(t) * G_\infty(\boldsymbol{x}'|\boldsymbol{x}^S, t)$$

$$- \int_{\boldsymbol{x}\in\partial\Omega} \mathcal{V}^B(\boldsymbol{x}|\boldsymbol{x}^S, t) * \partial_\nu G_\infty(\boldsymbol{x}|\boldsymbol{x}', t) dl(\boldsymbol{x}) \qquad (14.19)$$

for $\boldsymbol{x}' \in \Omega$, $\boldsymbol{x}^S \in \Omega$ with $t > 0$, where $G_\infty(\boldsymbol{x}'|\boldsymbol{x}^S, t)$ is the (causal) fundamental solution in $\mathbb{R}^2$ and the testing voltage distribution $\mathcal{V}^B(\boldsymbol{x}|\boldsymbol{x}^S, t)$ in the integration is available for $\boldsymbol{x} \in \partial\Gamma$ and $t > 0$ via TD-CIM. The first part of the right-hand side of Eq. (14.19) can be interpreted as the 'incident field' emanating from the testing source, while the second part represents the 'scattered field' due to the circuit's boundary $\partial\Omega$. The fact that for $\boldsymbol{x}' = \boldsymbol{x}^S$ only the former part becomes singular has motivated the choice $\boldsymbol{x}' = \boldsymbol{x}^C \notin \partial\Gamma$ in (14.18). This can be clarified in the complex-FD, with $\{s \in \mathbb{C}; \mathrm{Re}(s) > 0\}$ being its transformation parameter. To this end, without loss of generality, we consider $\hat{G}_\infty(\boldsymbol{x}|\boldsymbol{x}^S, s) = (1/2\pi)\mathrm{K}_0[sr(\boldsymbol{x}|\boldsymbol{x}^S)]$ in the contour integration along $\partial\Gamma$ and make use of the addition theorems for (modified) Bessel functions (see [111, Sec. 6.11] and [1, (9.6.3) and (9.6.4)]). This way leads to $\int_{\boldsymbol{x}\in\partial\Gamma} \mathrm{K}_0[sr(\boldsymbol{x}|\boldsymbol{x}^S)]dl = 2\pi\varrho\, \mathrm{I}_0(s\varrho/c)\mathrm{K}_0(s\varrho/c)$, for $\boldsymbol{x}^S$ approaching $\partial\Gamma$

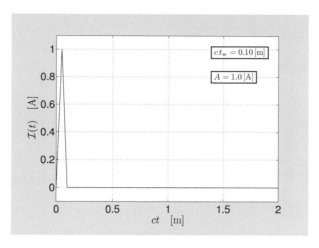

Figure 14.3: The triangular excitation signature.

from its exterior, which together with the limit $\lim_{s\varrho\downarrow 0} \mathrm{I}_0(s\varrho/c) = 1$ justifies $r(\boldsymbol{x}'|\boldsymbol{x}^S) = \varrho$ in (14.18).

The simplified Eq. (14.18) still represents an integral equation of the time-convolution type that must be solved for the unknown electric-current surface density $\partial J(t)$. To this end, we approximate the time convolution on its right-hand side using the trapezoidal rule with the uniform time step $\triangle t$ and get

$$\partial J(t) * \mathcal{V}^B(\boldsymbol{x}'|\boldsymbol{x}^S, t) \simeq \triangle t \sum_{k=1}^{N-1} \partial J(k\triangle t)\mathcal{V}^B(\boldsymbol{x}'|\boldsymbol{x}^S, t - k\triangle t) \quad (14.20)$$

where we use the fact that $\partial J(t)$ and $\mathcal{V}^B(\boldsymbol{x}'|\boldsymbol{x}^S, t)$ are zero for $t \leq 0$. Since the actual voltage response $\mathcal{V}^A(\boldsymbol{x}^S, t)$ is again easily available using TD-CIM and thanks to the fact that its time convolution with a suitably chosen $\mathcal{I}^B$ can be carried out analytically, the left-hand side of (14.18) is known. It is worth noting, in this respect, that since the testing voltage distribution is calculated using TD-CIM, the choice $\mathcal{I}^B(t) = \delta(t)$ that would filter out the convolution is not possible in general (cf. Eq. (3.14)). For the present calculations, the testing-source signature is chosen to be identical to the triangular pulse shape. Making use of (14.20) in (14.18) finally yields a relation from which $\partial J(k\triangle t)$ is successively attainable in a step-by-step manner for $\{t = k\triangle t, k = 1, \ldots, N\}$. The electric-current surface density evaluated along these lines is shown in Fig. 14.4.

With the electric-current surface density at our disposal we will use Eq. (14.16) to evaluate the voltage difference at the position of the

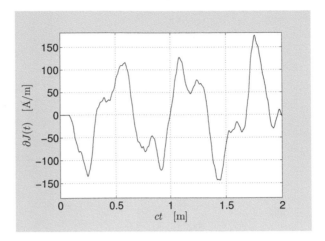

Figure 14.4: Electric-current surface density on the pin's surface.

voltage probe. To this end, we employ the approximation from the previous step and write

$$\mathcal{I}^B(t) * \mathcal{V}^\Delta(\boldsymbol{x}^S, t) \simeq 2\pi\varrho\,\partial J(t) * \mathcal{V}^B(\boldsymbol{x}'|\boldsymbol{x}^S, t) \qquad (14.21)$$

with $\boldsymbol{x}' = \boldsymbol{x}^C - [0, \varrho]$. Since $\boldsymbol{x}' \neq \boldsymbol{x}^S$, the evaluation of the testing-voltage response $\mathcal{V}^B(\boldsymbol{x}'|\boldsymbol{x}^S, t)$ does not present any difficulties and can be easily calculated using TD-CIM. In the present calculations, the testing source signature $\mathcal{I}^B$ is chosen, again, to be the (unit) triangular one. Once the time convolution with on the right-hand side of Eq. (14.21) is carried out, the remaining task is to perform the operation of deconvolution to recover the pulsed voltage difference. For a suitably chosen $\mathcal{I}^B$, this can be done directly by processing the pulses. Realizing the fact that the chosen triangular pulse can be expressed as the self-convolution of the corresponding rectangular pulse, we may use, for example, the following two-step procedure, symbolically

$$\mathcal{R}(t) = (t_\mathrm{w}/2)\partial_t\mathcal{F}(t) + \mathcal{R}(t - T_\mathrm{w}/2) \qquad (14.22)$$
$$\mathcal{V}^\Delta(t) = \partial_t\mathcal{R}(t) + \mathcal{V}^\Delta(t - t_\mathrm{w}/2) \qquad (14.23)$$

in which $\mathcal{F}(t)$ denotes the (known) right-hand side of Eq. (14.21) and $\mathcal{R}(t)$ corresponds to the time-convolution of the pulsed voltage difference $\mathcal{V}^\Delta(t)$ with the rectangular pulse. For an alternative closed-form deconvolution algorithm we refer the reader to Eq. (13.17). The resulting pulsed voltage difference is given in Fig. 14.5a. As the final step, the actual voltage response $\tilde{\mathcal{V}}^A$ is found from $\mathcal{V}^\Delta + \mathcal{V}^A$, where the pulsed voltage response of the circuit without the PIN is found using TD-CIM. In order to validate the introduced methodology, the planar circuit from

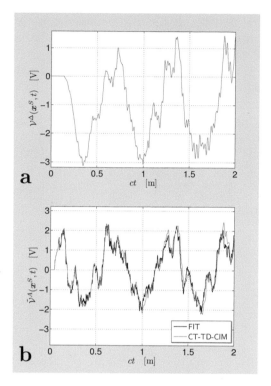

Figure 14.5: Pulsed voltage responses at the position of the PROBE. (a) Pulsed voltage difference; (b) actual pulsed voltage response.

Fig. 14.2 has also been analysed using the FIT as implemented in CST Microwave Studio®. The corresponding results, evaluated using the TD compensation theorem in combination with TD-CIM (i.e. CT-TD-CIM) and using FIT, are shown in Fig. 14.5b. Evidently, the results correlate very well. As has been numerically tested, the observable discrepancies at the late-time part of the voltage response can be reduced by refining the space-time discretization of the computational model. Such numerical experiments provide evidence for the convergence of the proposed solution.

14.6   Conclusions

Starting from the generic reciprocity theorem of the time-convolution type, the TD compensation theorems concerning arbitrarily-shaped inclusions with dielectric and conductive properties in planar circuits have been derived. Applications of the theorems to the evaluation of the impact of PEC and penetrable inclusions have been discussed with regard to the computational methodology. The results have then been applied to analyse the pulsed voltage response of an irregularly-shaped planar circuit with a shorting PEC pin. The obtained results have been successfully validated using (three-dimensional) FIT.

The introduced TD compensation theorems furnish not only efficient computational methodologies for analyzing various planar structures but in addition, and perhaps more notably, they provide useful physical insights into the impact of inclusions on the relevant pulsed EM field transfer.

Chapter 15

The Time-domain Compensation Contour Integral Method

The TD-CIM as formulated in Chapters 3 and 6 is an efficient TD IE technique for analyzing pulsed EM properties of arbitrarily shaped planar structures [127]. It will be demonstrated next that the high computational efficiency of this approach can further be increased through the concept of compensation, previously applied in Chapter 14 to analyse the effect of conductive and dielectric inclusions.

The methodology introduced in the present chapter finds its origin in the observation that the shape of a large class of irregularly-shaped planar structures can be formed upon a relatively small number of deformations of their circumscribing box. The boundary contours of the corresponding (two-dimensional) computational models then typically overlap along a significant part of the bounding rectangle. Now, owing to the fact that a rectangular resonator admits the TD ray-type (progressing) solution that can be readily evaluated within any prescribed accuracy (see Chapter 5), it would be computationally economical to take advantage of the known closed-form solution and formulate a TD contour IE whose boundary curve consists only of (the union of) those sections of the actual boundary contour that do not coincide with the rectangular one. Formulation and numerical solution of such an IE is the primary concern of this chapter, wherein the Time-Domain Compensation Contour Integral Method (TD-C^{2}IM) is introduced.

15.1 Problem formulation

We shall analyse the pulsed EM response of a planar circuit shown in Fig. 15.1. The problem is formulated using the reciprocity theorem of the time-convolution type (see [24, Sec. 28.2] and [133, Sec. 1.4.1]). In the following reciprocity-based analysis, we distinguish between the actual EM field state (state $\tilde{A}$) and the referential EM field state (state A) that corresponds to the circumscribing rectangular circuit. The conducting planes of the latter circuit occupy the rectangular domain Ω, whose boundary contour is denoted by $\partial\Omega$. The non-overlapping part of the actual boundary contour $\partial\tilde{\Omega}$ is then given as $\partial\Gamma = \partial\tilde{\Omega} \setminus (\partial\Omega \cap \partial\tilde{\Omega})$ (see Fig. 15.2). Both field states are, for the sake of simplicity, excited by a vertical excitation port whose (causal) action is accounted for by the electric-current volume density J_3^A, with $\text{supp}(J_3^A) \subset \tilde{\Omega}$, whose excitation pulse shape is $\tilde{\mathcal{I}}^A(t)$. The incremental change (not necessarily infinitesimally small) of the actual field quantities is defined, symbolically, as $\Delta A = \tilde{A} - A$. The actual field states will be interrelated with the testing field state (state $\tilde{B}$) that is activated by the concentrated source $J_3^B(\boldsymbol{x}, t) = \mathcal{I}^B(t)\delta(\boldsymbol{x} - \boldsymbol{x}^S)$. The actual and testing field states do satisfy the magnetic-wall boundary conditions (cf. Eq. (2.6))

$$\lim_{\delta\downarrow 0} \boldsymbol{\nu}(\boldsymbol{x}) \cdot \partial\tilde{\boldsymbol{J}}^A(\boldsymbol{x} + \delta\boldsymbol{\nu}, t) = 0 \text{ for all } \boldsymbol{x} \in \partial\tilde{\Omega} \qquad (15.1)$$

$$\lim_{\delta\downarrow 0} \boldsymbol{\nu}(\boldsymbol{x}) \cdot \partial\boldsymbol{J}^A(\boldsymbol{x} + \delta\boldsymbol{\nu}, t) = 0 \text{ for all } \boldsymbol{x} \in \partial\Omega \qquad (15.2)$$

$$\lim_{\delta\downarrow 0} \boldsymbol{\nu}(\boldsymbol{x}) \cdot \partial\boldsymbol{J}^B(\boldsymbol{x} + \delta\boldsymbol{\nu}|\boldsymbol{x}^S, t) = 0 \text{ for all } \boldsymbol{x} \in \partial\Omega \qquad (15.3)$$

where $\partial\boldsymbol{J}$ is the electric-current surface density that is associated with the corresponding magnetic-field strength $\boldsymbol{H} = \boldsymbol{H}(\boldsymbol{x}, t)$ according to $\partial\boldsymbol{J} = -\boldsymbol{i}_3 \times \boldsymbol{H}$ and $(\boldsymbol{x}|\boldsymbol{x}^S, t)$ indicates that the testing field is observed at $(\boldsymbol{x}, t)$ and excited by the causal point source localized at $\boldsymbol{x}^S$. Finally, recall that the thin-slab assumption allows us to define the pulsed voltage response $\mathcal{V} = -dE_3$ using the corresponding electric-field strength $E_3 = E_3(\boldsymbol{x}, t)$ (see Sec. 2.1).

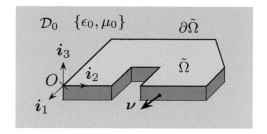

Figure 15.1: Irregularly-shaped planar circuit.

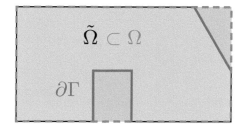

Figure 15.2: The rectangular domain Ω, its subset $\tilde{\Omega}$ that corresponds to the computational domain and 'the non-overlapping part' of the actual bounding contour $\partial\Gamma$.

In the first step, the reciprocity theorem is applied to ΔA and B states and to the irregularly-shaped domain $\tilde{\Omega}$. The resulting interaction quantity is with the help of Eqs. (15.1)–(15.3) written as (cf. Eqs. (2.29) and (14.4))

$$\int_{\boldsymbol{x}'\in\partial\Gamma} \mathcal{V}^B(\boldsymbol{x}'|\boldsymbol{x}^S,t) * \boldsymbol{\nu}(\boldsymbol{x}') \cdot \partial\boldsymbol{J}^A(\boldsymbol{x}',t)\mathrm{d}l(\boldsymbol{x}')$$

$$+\int_{\boldsymbol{x}'\in\partial\Gamma} \mathcal{V}^\Delta(\boldsymbol{x}',t) * \boldsymbol{\nu}(\boldsymbol{x}') \cdot \partial\boldsymbol{J}^B(\boldsymbol{x}'|\boldsymbol{x}^S,t)\mathrm{d}l(\boldsymbol{x}')$$

$$= \chi_{\tilde{\Omega}}(\boldsymbol{x}^S) \int_{\boldsymbol{x}'\in\tilde{\Omega}} \mathcal{V}^\Delta(\boldsymbol{x}',t) * J_3^B(\boldsymbol{x}'|\boldsymbol{x}^S,t)\mathrm{d}A(\boldsymbol{x}') \qquad (15.4)$$

for all $\boldsymbol{x}^S \in \Omega$ and $t > 0$. Secondly, the reciprocity theorem is applied to ΔA and B and to $\Omega \setminus \tilde{\Omega}$. Taking into account the orientation of the normal vector and the explicit-type boundary conditions (15.2) and

(15.3) we arrive at

$$\int_{\boldsymbol{x}'\in\partial\Gamma} \mathcal{V}^B(\boldsymbol{x}'|\boldsymbol{x}^S,t) * \boldsymbol{\nu}(\boldsymbol{x}') \cdot \partial\boldsymbol{J}^A(\boldsymbol{x}',t)\mathrm{d}l(\boldsymbol{x}')$$

$$-\int_{\boldsymbol{x}'\in\partial\Gamma} \mathcal{V}^A(\boldsymbol{x}',t) * \boldsymbol{\nu}(\boldsymbol{x}') \cdot \partial\boldsymbol{J}^B(\boldsymbol{x}'|\boldsymbol{x}^S,t)\mathrm{d}l(\boldsymbol{x}')$$

$$= \chi_{\Omega\setminus\tilde{\Omega}}(\boldsymbol{x}^S) \int_{\boldsymbol{x}'\in\tilde{\Omega}} \mathcal{V}^A(\boldsymbol{x}',t) * J_3^B(\boldsymbol{x}'|\boldsymbol{x}^S,t)\mathrm{d}A(\boldsymbol{x}') \qquad (15.5)$$

for all $\boldsymbol{x}^S \in \Omega$ and $t > 0$. Combination of Eqs. (15.4) and (15.5) results in the following TD contour IE, which is the point of departure for the TD-C²IM formulated in this chapter, that is

$$\tfrac{1}{2}\mathcal{I}^B(t) * \tilde{\mathcal{V}}^A(\boldsymbol{x}^S,t) - \int_{\boldsymbol{x}'\in\partial\Gamma} \tilde{\mathcal{V}}^A(\boldsymbol{x}',t) * \boldsymbol{\nu}(\boldsymbol{x}') \cdot \partial\boldsymbol{J}^B(\boldsymbol{x}'|\boldsymbol{x}^S,t)\mathrm{d}l(\boldsymbol{x}')$$

$$= \mathcal{I}^B(t) * \mathcal{V}^A(\boldsymbol{x}^S,t) \qquad (15.6)$$

for all $\boldsymbol{x}^S \in \partial\Gamma$ and $t > 0$. It should be emphasized that the solution domain of the resulting IE extends over the 'compensation boundary contour' $\partial\Gamma \subset \partial\tilde{\Omega}$ only (see Fig. 15.2), whose length depends on the 'similarity' of the actual circuit's shape $\partial\tilde{\Omega}$ with the corresponding circumscribed rectangle. Owing to the fact that A and B field states are associated with the rectangular circuit whose response is readily available via the TD ray-type (progressing) expansion (see Chapter 5), the only unknown quantity in Eq. (15.6) is the actual voltage distribution $\tilde{\mathcal{V}}^A$ along $\partial\Gamma$. Upon combining Eqs. (15.4)–(15.5) again, the latter distribution can be applied to calculate the voltage difference according to

$$\mathcal{I}^B(t) * \tilde{\mathcal{V}}^\Delta(\boldsymbol{x}^S,t)$$

$$= \int_{\boldsymbol{x}'\in\partial\Gamma} \tilde{\mathcal{V}}^A(\boldsymbol{x}',t) * \boldsymbol{\nu}(\boldsymbol{x}') \cdot \partial\boldsymbol{J}^B(\boldsymbol{x}'|\boldsymbol{x}^S,t)\mathrm{d}l(\boldsymbol{x}') \qquad (15.7)$$

for all $\boldsymbol{x}^S \in \tilde{\Omega}$ and $t > 0$. Since the pulsed voltage response of the rectangular circuit $\mathcal{V}^A$ is known, the desired actual voltage response in $\tilde{\Omega}$ directly follows as $\tilde{\mathcal{V}}^A = \mathcal{V}^A + \mathcal{V}^\Delta$.

15.2   Problem solution

The solution procedure largely follows the strategy applied in Chapter 3. In short, the boundary contour $\partial\Gamma$ is approximated by NS line segments $\partial\Gamma \simeq \cup_{m=1}^{NS}\Delta\Gamma^{[m]}$ and the solution is carried out along the uniform time axis in the bounded time window of observation $\mathcal{T} = \{t \in \mathbb{R}; t = k\triangle t, \triangle t > 0, k = 1, 2, \ldots, NT\} \subset \{t \in \mathbb{R}; t > 0\}$. Secondly, the actual voltage distribution is in space-time expanded in a set of piecewise-linear basis functions (cf. Eq. (3.4))

$$\tilde{\mathcal{V}}^A(\boldsymbol{x}, t) \simeq \sum_{m=1}^{NQ} \sum_{k=1}^{NT} \tilde{v}_{[k]}^{[A;m]} T^{[m]}(\boldsymbol{x}) T_{[k]}(t) \tag{15.8}$$

where $\tilde{v}_{[k]}^{[A;m]}$ are unknown coefficients, $T^{[m]}(\boldsymbol{x})$ and $T_{[k]}(t)$ are the spatial and temporal triangular functions (see Eqs. (3.5) and (3.6)), respectively, and NQ is the number of nodes along the approximated $\partial\Gamma$ contour. Note that $\partial\Gamma$ does not necessarily have to be closed and, hence, $NQ \neq NS$, in general. The integral equation is next 'weighted' by the piecewise linear spatial function $T^{[S]}(\boldsymbol{x}^S)$ and integrated along $\partial\Gamma$. This way leads to the following system of equations that is solvable in an updating manner according to (cf. Eq. (3.8))

$$\left(\boldsymbol{I} - \boldsymbol{Q}_{[0]}\right) \cdot \tilde{\mathcal{V}}_{[p]}^A = \sum_{k=1}^{p-1} \boldsymbol{Q}_{[p-k]} \cdot \tilde{\mathcal{V}}_{[p]}^A + \boldsymbol{F}_{[p]} \tag{15.9}$$

for $p = \{1, \ldots, NT\}$, successively. Here, $\tilde{\mathcal{V}}_{[p]}^A$ is a 2D-array of $[NQ \times NT]$ unknown coefficients at $t = p\triangle t$ and $\boldsymbol{I}$ is a three-diagonal $[NQ \times NQ]$ array whose elements for a *closed* 'compensation contour' $\partial\Gamma$ are described as (cf. Eq. (3.9))

$$(\boldsymbol{I})_{S,m} = \left(\triangle\Omega^{[S-1]}/12\right)\delta_{S-1,m}$$
$$+ \left(\triangle\Omega^{[S-1]}/6 + \triangle\Omega^{[S]}/6\right)\delta_{S,m} + \left(\triangle\Omega^{[S]}/12\right)\delta_{S+1,m} \tag{15.10}$$

Next, $\boldsymbol{Q}_{[p-k]}$ is a time-dependent $[NQ \times NQ \times NT]$ 3D-array whose elements are given as (cf. Eq. (3.10))

$$(\boldsymbol{Q}_{[p-k]})_{S,m} = \frac{1}{2\pi c\triangle t} \int_{\boldsymbol{x}^T \in \partial\Gamma} T^{[S]}(\boldsymbol{x}^T) \int_{\boldsymbol{x} \in \partial\Gamma} T^{[m]}(\boldsymbol{x})$$
$$\sum_{u \in U} \sum_{v \in V} \Theta[\boldsymbol{x}|\boldsymbol{x}^T(u, v), (p - k)\triangle t] \mathrm{d}l(\boldsymbol{x}) \mathrm{d}l(\boldsymbol{x}^T) \tag{15.11}$$

for all $S = \{1, \cdots, NQ\}$, $m = \{1, \cdots, NQ\}$ and $t \in \mathcal{T}$, where

$$\Theta[\boldsymbol{x}|\boldsymbol{x}^T(u, v), t]$$
$$= \Psi\{r^{++}[\boldsymbol{x}|\boldsymbol{x}^T(u, v)], t\} \cos\{\theta^{++}[\boldsymbol{x}|\boldsymbol{x}^T(u, v)]\}$$
$$+ \Psi\{r^{+-}[\boldsymbol{x}|\boldsymbol{x}^T(u, v)], t\} \cos\{\theta^{+-}[\boldsymbol{x}|\boldsymbol{x}^T(u, v)]\}$$
$$+ \Psi\{r^{-+}[\boldsymbol{x}|\boldsymbol{x}^T(u, v)], t\} \cos\{\theta^{-+}[\boldsymbol{x}|\boldsymbol{x}^T(u, v)]\}$$
$$+ \Psi\{r^{--}[\boldsymbol{x}|\boldsymbol{x}^T(u, v)], t\} \cos\{\theta^{--}[\boldsymbol{x}|\boldsymbol{x}^T(u, v)]\} \tag{15.12}$$

in which r and $\cos(\theta)$, supplemented with the corresponding $\pm$ symbol, follow from $r = (\boldsymbol{d} \cdot \boldsymbol{d})^{1/2} > 0$ and $\cos(\theta) = \boldsymbol{\nu}(\boldsymbol{x}) \cdot \boldsymbol{d} / (\boldsymbol{d} \cdot \boldsymbol{d})^{1/2}$, with

$$\boldsymbol{d}^{++} = [x_1 - x_1^T - 2uL, x_2 - x_2^T - 2vW] \tag{15.13}$$

$$\boldsymbol{d}^{+-} = [x_1 - x_1^T - 2uL, x_2 + x_2^T - 2vW] \tag{15.14}$$

$$\boldsymbol{d}^{-+} = [x_1 + x_1^T - 2uL, x_2 - x_2^T - 2vW] \tag{15.15}$$

$$\boldsymbol{d}^{--} = [x_1 + x_1^T - 2uL, x_2 + x_2^T - 2vW] \tag{15.16}$$

Finally, the space-time function in Eq. (15.12) is for a loss-free circuit given in Eq. (3.12) and its generalisations, accounting for relaxation phenomena, can be evaluated via Eqs. (6.19) and (6.24). The last term on the right-hand side of Eq. (15.9) is an $[NQ \times NT]$ 2D-array $\boldsymbol{F}_{[p]}$ whose elements read

$$(\boldsymbol{F}_{[p]})_S = \int_{\boldsymbol{x}^T \in \partial \Gamma} T^{[S]}(\boldsymbol{x}^T) \mathcal{V}^A(\boldsymbol{x}^T, p\triangle t) \mathrm{d}l(\boldsymbol{x}^T) \tag{15.17}$$

for all $S = \{1, \cdots, NQ\}$ and all $t \in \mathcal{T}$ and $\mathcal{V}^A$ is given by the ray-type expansion closely described in Chapter 5. A relatively rough, yet frequently sufficient (midpoint) approximation of Eq. (15.17) is given as

$$\begin{aligned}(\boldsymbol{F}_{[p]})_S \simeq &\tfrac{1}{2} \triangle \Gamma^{[S-1]} \mathcal{V}^A(\boldsymbol{x}^{[c;S-1]}, p\triangle t) \\ &+ \tfrac{1}{2} \triangle \Gamma^{[S]} \mathcal{V}^A(\boldsymbol{x}^{[c;S]}, p\triangle t)\end{aligned} \tag{15.18}$$

where $\boldsymbol{x}^{[c;S]} = (\boldsymbol{x}^{[S]} + \boldsymbol{x}^{[S+1]})/2$ is the center of the S-th segment. Due to the property of causality that is preserved throughout the proposed solution, the summations in Eq. (15.11) always contain a *finite* number of successively arriving ray-type constituents. As their number increases with the length of the time window, the solution procedure is expected to be computationally efficient, particularly for calculating the early-time response.

With the voltage nodal coefficients at our disposal, the midpoint integration rule allows us to evaluate the voltage difference from Eq. (15.7) according to

$$\begin{aligned}\tilde{\mathcal{V}}^A(\boldsymbol{x}^S, p\triangle t) \simeq &\frac{1}{2\pi c \triangle t} \sum_{m=1}^{NS} \triangle \Gamma^{[m]} \sum_{k=1}^{p} \frac{\tilde{v}_{[k]}^{[A;m]} + \tilde{v}_{[k]}^{[A;m+1]}}{2} \\ &\sum_{u \in U'} \sum_{v \in V'} \Theta[\boldsymbol{x}^{[c;n]} | \boldsymbol{x}^S(u, v), (p - k)\triangle t]\end{aligned} \tag{15.19}$$

for all $\boldsymbol{x}^S \in \tilde{\Omega}$ and $t \in \mathcal{T}$, where $(\tilde{v}_{[k]}^{[A;m]} + \tilde{v}_{[k]}^{[A;m+1]})/2$ is the arithmetic average of the nodal coefficients that correspond to the m-th line segment. Again, the total number of the ray-type constituents is always finite. As for the implementation aspects of Eq. (15.19), the multiplications over the temporal index can be carried out in a straightforward manner using `conv` function as implemented in MATLAB®, for example. Furthermore, it is important to realize that the introduced formulation does not require the elaborate handling of the singularity that occurs in the basic TD-CIM formulation (see Sec. 2.1.4). Since $\mathcal{V}^\Delta$ is source-free in $\tilde{\Omega}$, the only difficulty in this respect is the direct ray-type constituent of $\mathcal{V}^A$ when calculating the self-response of the circuit. This term, however, can be handled by considering a finite radius of the port, say $\varrho > 0$, and making use of the addition theorems for (modified) Bessel functions (see [111, Sec. 6.11] and [1, (9.6.3) and (9.6.4)]) in the complex-FD. Along these lines we get the following expression for the direct-ray constituent

$$(\mu_0 d/2\pi)\partial_t \tilde{\mathcal{I}}^A(t) * (t^2 - \varrho^2/c^2)^{-1/2}\mathrm{H}(t - \varrho/c) \tag{15.20}$$

where we have assumed that $\varrho \ll ct_\mathrm{w}$. For a significant number of the excitation pulse shapes (e.g. Eq. (F.6)), the time convolution in Eq. (15.20) can be calculated analytically.

15.3 Numerical results

In this section we analyse the irregularly-shaped parallel-plane circuit shown in Fig. 15.3. Its thickness is $d = 1.50\,[\mathrm{mm}]$ and the EM properties of the slab are described by $\epsilon = 4.50\epsilon_0$ and $\mu = \mu_0$. The corresponding circumscribed rectangular circuit (see Fig. 5.1) occupies the box specified by $\{0 \le x_1 \le L, 0 \le x_2 \le W, 0 \le x_3 \le d\}$ with $L = 50.0\,[\mathrm{mm}]$ and $W = 75.0\,[\mathrm{mm}]$.

The analysed circuit is excited via the electric-current port whose center is placed at $\boldsymbol{x}^C = \{3L/4, W/4\}$. Its radius is $\varrho = 0.10\,[\mathrm{mm}]$ (PORT). The excitation electric-current pulse has the triangular shape that is described by Eq. (5.24), where we take $A = 1.0\,[\mathrm{A}]$ and $ct_\mathrm{w} = L$. The pulsed voltage response of the structure is subsequently observed at $\boldsymbol{x}^S = \{L/4, W/2\}$ (PROBE) within the time window $\mathcal{T} = \{0 \le ct \le 10L\}$ with the (scaled) time step $c\triangle t = L/50$. The 'compensation contour' $\partial\Gamma$ is divided into the 12 line segments, such that $\max_m |\triangle\Gamma^{[m]}| < ct_\mathrm{w}/8$ (see Fig. 15.3).

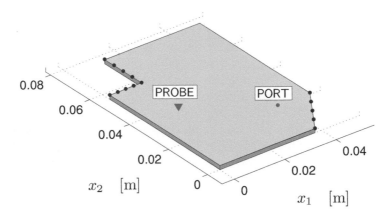

Figure 15.3: Analysed parallel-plate circuit and the discretization nodes along its $\partial\Gamma$ contour.

In the first step, we solve the integral equation (15.6) for the actual voltage distribution $\tilde{\mathcal{V}}^A(\boldsymbol{x}^S, t)$ with $\boldsymbol{x}^S \in \partial\Gamma$ and $t \in \mathcal{T}$. For validation purposes, the voltage response at the corner point $\boldsymbol{x}^S = \{L/4, 3W/4\}$ has also been evaluated with the aid of the FIT as implemented in CST Microwave Studio®. Figure 15.4a shows a good correspondence with the alternative numerical technique, thereby validating the step-by-step updating solution procedure introduced in Sec. 15.2.

Subsequently, the evaluated voltage coefficients at the nodal points are substituted in Eq. (15.19), which leads to the pulsed voltage difference at the position of the probe at $\boldsymbol{x}^S = \{L/4, W/2\}$. Since the response of the corresponding rectangular circuit $\mathcal{V}^A(\boldsymbol{x}^S, t)$ is known, the actual voltage immediately follows. The resulting voltage pulses are shown in Fig. 15.4b. Again, the resulting pulse shapes evaluated using the TD-C²IM and the referential FIT correlate very well. The former method, however, has managed to calculate the pulse shapes with only 12 discretization elements, thereby introducing extremely high computational savings of several orders of magnitude with respect to the reference.

Finally, in order to validate the suggested handling of the singularity, we calculate the self-response of the analysed circuit using Eqs. (15.9) and (15.19) with (15.20). Figure 15.4c shows the results that correlate very well.

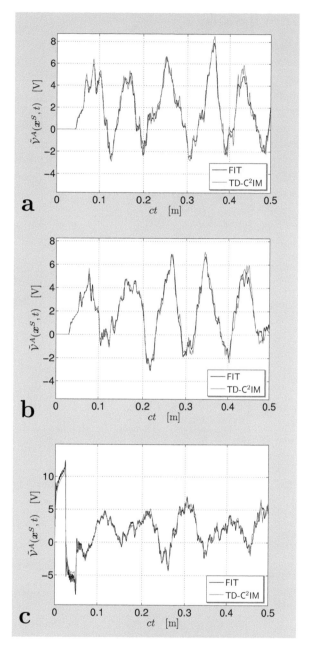

Figure 15.4: The actual pulsed voltage response at (a) $x^S = \{L/4, 3W/4\} \in \partial\Gamma$; (b) $x^S = \{L/4, W/2\} \in \tilde{\Omega}$ (PROBE); (c) $x^S = \{3L/4, W/4\} \in \tilde{\Omega}$ (PORT).

15.4 Conclusions

In this chapter we have introduced a novel TD IE for analyzing planar circuits of arbitrary shape. The formulation of the computational technique is based on a systematic use of the TD reciprocity theorem in combination with the ray-type, closed-form TD solution for a rectangular planar circuit being taken as reference. It has been demonstrated that this compensation-based formulation may reduce the spatial domain of the starting IE to a small section of the actual bounding contour and, hence, drastically decrease the number of unknowns in the resulting system of equations. Finally, the TD-C^{2}IM has been validated with the help of (three-dimensional) FIT.

Chapter 16

Modeling of Shorting Via Structures

Shorting vias in parallel-plane structures constitute important functional components that have found a wealth of applications in suppressing radiated EMI from PCBs [142], in designing microwave filters [44], reconfigurable [76] and textile antennas [145], frequency-selective surfaces [141] and many others micro- and millimeter-wave components (see e.g. [10]).

As to their modeling, a number of efficient numerical techniques have been successfully applied. Apart from standard direct-discretization methods (see [144], for example), the concepts of periodic frill array [42] and the boundary integral-resonant mode expansion method [11, 13] have proved to be computationally efficient. The corresponding TD analysis is almost exclusively limited to applications of the finite-difference time-domain method (see [62], for instance). In this respect, it seems that no dedicated TD integral-equation technique has been proposed so far. An exception in this category is the TD compensation-based computational procedure briefly introduced in Sec. 14.5. The latter solution methodology, however, applies directly to a single circular PEC inclusion only. This observation motivates to formulate an extension of the method that is capable of analyzing the impact of a set of via structures on the pulsed EM transfer in parallel-plate environments. Accordingly, such an extension is the main focus of this chapter.

In the presented methodology, each shorting via is viewed as an inclusion that disturbs the circuit from its referential operation in the absence of vias. The corresponding solution procedure consists of two steps. In the first one, a time-convolution integral equation of the first kind is formulated using the relevant TD compensation theorem. Under the assumption that the diameter of via structures is small, with respect to the spatial support of the excitation pulse, the integral equation is cast into a system of time-convolution integral equations that is subsequently solved with the aid of the standard trapezoidal rule. This way leads to an updating step-by-step scheme that yields the unknown vector of electric currents associated with via structures.

The second step can be viewed as a mere calculation of the impact of via inclusions. This step is readily carried out, with the electric-current pulses at our disposal, using the compensation theorem, again. Since such a modular approach makes it possible to assess the impact of each via separately, the introduced methodology lends itself to efficient optimization procedures.

16.1 Problem definition

The problem configuration shown in Fig. 16.1 can be understood as a special case of the one described in Sec. 14.1. It consists of an arbitrarily-shaped parallel-plane structure that contains NV shorting PEC circular cylinders connecting the upper and bottom PEC plates. The PEC plates occupy domain Ω that is bounded by a closed contour denoted by $\partial\Omega$. The n-th PEC via occupies a circular domain $\Gamma^{[n]}$ whose bounding contour is denoted by $\partial\Gamma^{[n]}$, $n = \{1, \ldots, NV\}$. Its center is placed at $\boldsymbol{x}^{[n]}$ and its radius is $\varrho^{[n]}$. The relevant unions are denoted by $\Gamma = \cup_{n=1}^{NV}\Gamma^{[n]}$ and $\partial\Gamma = \cup_{n=1}^{NV}\partial\Gamma^{[n]}$. Finally, the outer normal vector is denoted by $\boldsymbol{\nu}$. The structure is placed in the homogeneous isotropic embedding that is described by its scalar permittivity ϵ_0 and permeability μ_0.

In accordance with Chapter 14, we distinguish between the pulsed voltage response $\mathcal{V}^A$ that would be excited in the absence of the via inclusions (state A) and the actual voltage response $\tilde{\mathcal{V}}^A$ that accounts for their presence (state $\tilde{A}$). Accordingly, the incremental change (not necessarily infinitesimally small) of the actual voltage response is defined as $\mathcal{V}^\Delta = \tilde{\mathcal{V}}^A - \mathcal{V}^A$. The actual voltage response is interrelated with the testing one denoted by $\mathcal{V}^B$ (state B) that is associated with its excitation electric current $\mathcal{I}^B$.

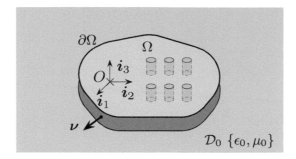

Figure 16.1: Planar circuit with shorting via structures.

16.2   Problem solution

The problem solution relies on the compensation theorem for perfectly conducting inclusions introduced in Sec. 14.3. In Sec. 14.5, the impact of a single shorting via is evaluated for validation purposes. The applied procedure, however, has proved to be prone to divergence for a higher number of PEC inclusions. A possible solution that eliminates errors arising from approximation of the superposition integral (see Eq. (14.19)) makes use the concept of 'virtual' ports [136]. In this approach, the solution procedure is carried out along the boundary contour consisting of $\partial\Omega$ and $\partial\Gamma$. This strategy that results directly in the testing-voltage response along the (discretized) periphery of each via is followed in the proposed solution.

The solution procedure starts with the first compensation theorem

$$
\mathcal{I}^B(t) * \left[\mathcal{V}^\Delta(\boldsymbol{x}^S,t)\chi_{\Omega\backslash\Gamma}(\boldsymbol{x}^S) - \mathcal{V}^A(\boldsymbol{x}^S,t)\chi_\Gamma(\boldsymbol{x}^S) \right]
$$
$$
= \int_{\boldsymbol{x}'\in\partial\Gamma} \mathcal{V}^B(\boldsymbol{x}'|\boldsymbol{x}^S,t) * \boldsymbol{\nu}(\boldsymbol{x}') \cdot \partial\tilde{\boldsymbol{J}}^A(\boldsymbol{x}',t)\mathrm{d}l(\boldsymbol{x}')
$$
$$
- \int_{\boldsymbol{x}'\in\partial\Gamma} \tilde{\mathcal{V}}^A(\boldsymbol{x}',t) * \boldsymbol{\nu}(\boldsymbol{x}') \cdot \partial\boldsymbol{J}^B(\boldsymbol{x}'|\boldsymbol{x}^S,t)\mathrm{d}l(\boldsymbol{x}') \quad \text{(14.9 revisited)}
$$

for all $\boldsymbol{x}^S \in \Omega$ and all $t > 0$, where $\boldsymbol{\nu} \cdot \partial\boldsymbol{J}$ represents (the normal component of) the electric-current surface density flowing on the vias' surface. The two-step solution procedure will be carried out along the discrete time axis in the bounded time window of observation $\mathcal{T} = \{t \in \mathbb{R}; t = n\triangle t, \triangle t > 0, n = 1, 2, \ldots, NT\} \subset \{t \in \mathbb{R}; t > 0\}$.

16.2.1 Solving the integral equation

Upon enforcing the explicit-type boundary condition $\lim_{\delta \downarrow 0} \tilde{\mathcal{V}}^A(\boldsymbol{x}^S +$ $\delta \boldsymbol{\nu}, t) = 0$ as $\delta \downarrow 0$ for all $\boldsymbol{x}^S \in \partial \Gamma$ and $t > 0$ on the PEC surface of each via, Eq. (14.9) takes the form of an integral equation of the first kind

$$\int_{\boldsymbol{x}' \in \partial \Gamma} \mathcal{V}^B(\boldsymbol{x}'|\boldsymbol{x}^S, t) * \boldsymbol{\nu}(\boldsymbol{x}') \cdot \partial \tilde{\boldsymbol{J}}^A(\boldsymbol{x}', t) \mathrm{d}l(\boldsymbol{x}')$$
$$= -\mathcal{I}^B(t) * \mathcal{V}^A(\boldsymbol{x}^S, t) \quad (14.15 \text{ revisited})$$

for $\boldsymbol{x}^S \in \partial \Gamma$ and $t > 0$. Now, instead of solving the integral equation, we shall focus on its 'spatially weighted' version, namely

$$\int_{\boldsymbol{x}^S \in \partial \Gamma} \mathrm{d}l(\boldsymbol{x}^S) \int_{\boldsymbol{x}' \in \partial \Gamma} \mathcal{V}^B(\boldsymbol{x}'|\boldsymbol{x}^S, t) * \boldsymbol{\nu}(\boldsymbol{x}') \cdot \partial \tilde{\boldsymbol{J}}^A(\boldsymbol{x}', t) \mathrm{d}l(\boldsymbol{x}')$$
$$= -\mathcal{I}^B(t) * \int_{\boldsymbol{x}^S \in \partial \Gamma} \mathcal{V}^A(\boldsymbol{x}^S, t) \mathrm{d}l(\boldsymbol{x}^S) \quad (16.1)$$

The latter integral relation is then solved in an approximate way. To this end, the vias' radius is assumed to be small, with respect to the spatial extent of the excitation pulse. Consequently, the electric-current surface density is approximately constant along each via's contour, which yields

$$\underline{\boldsymbol{\mathcal{V}}}^B(t) * \tilde{\boldsymbol{\mathcal{I}}}^A(t) \simeq -\boldsymbol{\mathcal{V}}^A(t) * \mathcal{I}^B(t) \quad (16.2)$$

where $\underline{\boldsymbol{\mathcal{V}}}^B(t)$ is an $[NV \times NV \times NT]$ 3D-array, whose elements follow from

$$\left[\underline{\boldsymbol{\mathcal{V}}}^B(t)\right]_{m,n} = \frac{1}{2\pi \varrho^{[m]}} \int_{\boldsymbol{x}^S \in \partial \Gamma^{[m]}} \mathcal{V}^B(\boldsymbol{x}^S|\boldsymbol{x}^{[n]}, t) \mathrm{d}l(\boldsymbol{x}^S) \quad (16.3)$$

Next, $\tilde{\boldsymbol{\mathcal{I}}}^A(t)$ is an (unknown) $[NV \times NT]$ 2D-array, whose n-th element is the electric current associated with the n-th via. The left-hand side of Eq. (16.2) implicitly includes the standard product of $[NV \times NV]$ and $[NV \times 1]$ arrays with the time parameter $t \in \mathcal{T}$. Finally, $\boldsymbol{\mathcal{V}}^A(t)$ is an $[NV \times NT]$ 2D-array of the actual-voltage response in absence of vias. Its m-th element is related to the voltage response along $\partial \Gamma^{[m]}$ according to

$$\left[\boldsymbol{\mathcal{V}}^A(t)\right]_m = \frac{1}{2\pi \varrho^{[m]}} \int_{\boldsymbol{x}^S \in \partial \Gamma^{[m]}} \mathcal{V}^A(\boldsymbol{x}^S, t) \mathrm{d}l(\boldsymbol{x}^S) \quad (16.4)$$

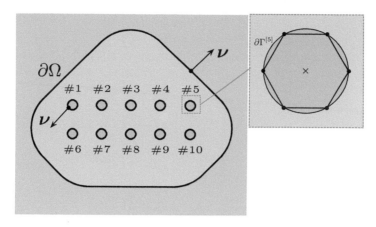

Figure 16.2: The solution domain with a detailed via's rim approximation.

Owing to the piecewise linear (space-time) expansion employed in TD-CIM, the latter integrations along the approximated via's rim (see Fig. 16.2) can be carried out analytically, giving us

$$\left[\boldsymbol{\mathcal{V}}^A(t)\right]_m \simeq P^{-1}\sum_{p=1}^{P}\mathcal{V}^A(\boldsymbol{x}^{[m]}+\boldsymbol{t}^{[m]}(p),t) \tag{16.5}$$

$$\left[\underline{\boldsymbol{\mathcal{V}}}^B(t)\right]_{m,n} \simeq P^{-1}\sum_{p=1}^{P}\mathcal{V}^B[\boldsymbol{x}^{[m]}+\boldsymbol{t}^{[m]}(p)|\boldsymbol{x}^{[n]},t] \tag{16.6}$$

with $P \geq 3$ being the number of line segments along the via's rim and $\boldsymbol{t}^{[m]}(p) = \varrho^{[m]}\{cos(2\pi p/P), \sin(2\pi p/P)\}$. Accordingly, the testing-voltage 3D-array $\underline{\boldsymbol{\mathcal{V}}}^B(t)$ and the actual-voltage 2D-array $\boldsymbol{\mathcal{V}}^A(t)$ can be composed of (the average of) voltage nodal values that directly follow from TD-CIM. Finally, note that the testing-voltage array can be associated with the corresponding 'TD impedance matrix' via $\underline{\boldsymbol{\mathcal{V}}}^B(t) = \underline{\boldsymbol{\mathcal{Z}}}^B(t) * \boldsymbol{\mathcal{I}}^B(t)$ that is symmetrical for (reciprocal) media under consideration.

The system of equations (16.2) can be interpreted as a system of the convolution-type (Volterra) equations that can be solved with the aid of the trapezoidal rule according to

$$\underline{\boldsymbol{\mathcal{V}}}^B(t) * \tilde{\boldsymbol{\mathcal{I}}}^A(t)|_{t=k\triangle t} = \int_{\tau=0}^{k\triangle t}\underline{\boldsymbol{\mathcal{V}}}^B(k\triangle t - \tau)\,\tilde{\boldsymbol{\mathcal{I}}}^A(\tau)\mathrm{d}\tau$$

$$\simeq \triangle t\sum_{q=1}^{k-1}\underline{\boldsymbol{\mathcal{V}}}^B[(k-q)\triangle t]\,\tilde{\boldsymbol{\mathcal{I}}}^A(q\triangle t) \tag{16.7}$$

for $k = \{2, 3, \ldots\}$. In the latter approximation we have used the fact that $\underline{\mathcal{V}}^B(0) = \underline{0}$ and $\tilde{\mathcal{I}}^A(0) = 0$ for all $m, n = \{1, \ldots, NV\}$. Making use of the trapezoidal-rule approximation in Eq. (16.2), we may express the unknown electric-current via the following step-by-step updating scheme

$$\tilde{\mathcal{I}}^A(k\triangle t) = - \left[\underline{\mathcal{V}}^B(\triangle t) \right]^{-1} \cdot \left\{ \mathcal{V}^A(t) * \mathcal{I}^B(t)|_{t=(k+1)\triangle t} / \triangle t \right.$$

$$\left. + \sum_{q=1}^{k-1} \underline{\mathcal{V}}^B[(k - q + 1)\triangle t] \tilde{\mathcal{I}}^A(q\triangle t) \right\} \quad (16.8)$$

for $k = \{1, 2, \ldots\}$, successively, provided that $c\triangle t > \varrho^{[m]}$ for all $m = \{1, \ldots, NV\}$. Obviously, the matrix inverse in Eq. (16.8) is calculated only once, whatever k. Moreover, the time convolution on the right-hand side of Eq. (16.2) can be calculated analytically for a judicious choice of the testing electric current $\mathcal{I}^B(t)$ (see Appendix F, for example).

16.2.2 Evaluating the impact of shorting vias

Once the electric-current on the via structures is known, the compensation theorem (14.9) is applied again in order to evaluate the vias' impact on the actual-voltage distribution. To this end, we write

$$\mathcal{I}^B(t) * V^\triangle(x^S, t)$$

$$= \int_{x' \in \partial\Gamma} \mathcal{V}^B(x'|x^S, t) * \nu(x') \cdot \partial \tilde{J}^A(x', t) \mathrm{d}l(x') \quad (16.9)$$

for $x^S \in \Omega \backslash \Gamma$ and $t > 0$. Assuming, again, the constant electric-current distribution along each via's contour $\partial\Gamma^{[n]}$, Eq. (16.9) can be rewritten as

$$\mathcal{I}^B(t) * V^\triangle(x^S, t) \simeq \underline{\mathcal{V}}^B(t) * \tilde{\mathcal{I}}^A(t) \quad (16.10)$$

where $\underline{\mathcal{V}}^B(t)$ is an $[NT \times NV]$ 2D-array, whose elements are given as

$$\left[\underline{\mathcal{V}}^B(t) \right]_n = \frac{1}{2\pi \varrho^{[n]}} \int_{x' \in \partial\Gamma^{[n]}} \mathcal{V}^B(x'|x^S, t) \mathrm{d}l(x') \quad (16.11)$$

The right-hand side of Eq. (16.10) implicitly includes the standard product of $[1 \times NV]$ and $[NV \times 1]$ arrays with the time parameter $t \in \mathcal{T}$. Again, the integral in Eq. (16.11) can be carried out analytically, which makes it possible to express the elements of $\boldsymbol{\mathcal{V}}^B(t)$ in terms of nodal values along the contour encircling the n-th via, that is

$$\left[\boldsymbol{\mathcal{V}}^B(t) \right]_n \simeq P^{-1} \sum_{p=1}^{P} \mathcal{V}^B(\boldsymbol{x}^{[n]} + \boldsymbol{t}^{[n]}(p), t) \qquad (16.12)$$

with $P \geq 3$ and $\boldsymbol{t}^{[n]}(p) = \varrho^{[n]}\{cos(2\pi p/P), sin(2\pi p/P)\}$. With the testing-voltage and actual electric-current 2D-arrays at our disposal, we may evaluate the time-convolution on the right-hand side of Eq. (16.10). Similarly to Eq. (16.7), this can be readily done using the trapezoidal rule along the uniform time grid, i.e.

$$\boldsymbol{\mathcal{V}}^B(t) * \tilde{\boldsymbol{\mathcal{I}}}^A(t)|_{t=k\triangle t} \simeq \triangle t \sum_{q=1}^{k-1} \boldsymbol{\mathcal{V}}^B[(k-q)\triangle t] \tilde{\boldsymbol{\mathcal{I}}}^A(q\triangle t) \qquad (16.13)$$

for $k = \{2, 3, \ldots\}$. In the final step, the voltage difference is evaluated using the deconvolution algorithm applied in Sec. 11.7 or, alternatively, the one used in Sec. 9.3. This procedure results in the pulsed-voltage response $\mathcal{V}^\triangle(\boldsymbol{x}^S, t)$ in $\boldsymbol{x}^S \in \Omega \setminus \Gamma$ and $t \in \mathcal{T}$. The desired actual voltage response then directly follows from $\tilde{\mathcal{V}}^A = \mathcal{V}^A + \mathcal{V}^\triangle$.

16.3   Numerical results

In this section, we analyse the impact of shorting via structures of radius $\varrho = 0.50$ [mm] on the pulsed-EM field transfer within two parallel-plane circuits. Their thickness is $d = 1.50$ [mm] and their relative electric permittivity is $\epsilon_r = 4.20$. The corresponding EM wave speed in the slab is $c = c_0/\sqrt{\epsilon_r}$.

The calculated pulsed-voltage responses are excited via a vertical electric-current port whose radius is $\varrho = 0.50$ [mm]. For the sake of simplicity, the excitation electric-current pulse shape is taken to be identical to the testing one that is described by Eq. (F.6). Here, we take $A = 1.0$ [A] and $ct_w = 0.10$ [m] (see Fig. 3.3). For the given constitution parameters, this approximately corresponds to $t_w \simeq 0.68$ [ns]. The pulsed voltage response is observed within the bounded time window of observation $\mathcal{T} = \{0 \leq ct \leq 3.0\}$ [m] with $c\triangle t = 10\varrho = 5.0$ [mm].

16.3.1 Irregularly-shaped planar circuit

The analysed irregularly-shaped planar circuit with its shorting via structures is shown in Fig. 16.3. In this configuration, the via structures are uniformly distributed along a circle having its center at $x^C = \{40, 50\}$ [mm] and radius $\varrho^C = 10.0$ [mm]. The vias are placed at $x^C + \varrho^C \{cos(n\pi/3), sin(n\pi/3)\}$ for $n = \{1, \ldots, 6\}$. The excitation port (PORT) is placed at $\{40, 20\}$ [mm] and the pulsed-voltage response is observed at $x^S = \{40, 80\}$ [mm] (PROBE). The circular rim of the excitation port, the probe and the shorting vias is approximated by $P = 8$ line segments. The maximum length of the line segments along the circuit's periphery is equal to $ct_w/20$.

The resulting electric currents that are associated with the vias for $n = \{1, 2, 3\}$, for example, are plotted in Fig. 16.4. Subsequently, the 'scattered voltage' due to all the via structures is shown in Fig. 16.5a. This pulse, in fact, represents the impact of the vias on the signal transfer between the exciting port and the probe. Combining the scattered field with the pulsed-voltage response in absence of the vias (see Fig. 16.5b) we get the sought response in the presence of the vias. The results, as evaluated using the introduced TD-CIM-based approach and the referential FIT, are shown in Fig. 16.5c. Finally, note that, in accordance with the relevant real-FD theory [88], the shorting via structures have visibly raised the low-frequency contents of the response $\mathcal{V}^A$ to higher frequencies.

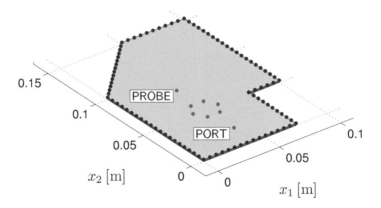

Figure 16.3: The analysed irregularly-shaped circuit with shorting via structures.

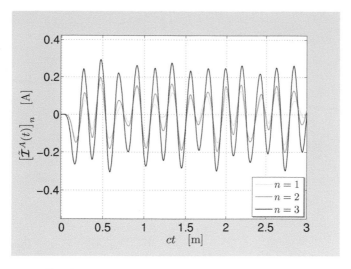

Figure 16.4: The electric currents induced on selected shorting via structures.

16.3.2   Rectangular planar circuit

The next analysed planar circuit with its shorting via structures is shown in Fig. 16.6. Its conducting planes occupy a rectangular domain $\{0 \leq x_1 \leq L, 0 \leq x_2 \leq W\}$, with $L = 50\,[\text{mm}]$ and $W = 75\,[\text{mm}]$. Three via structures are placed at $\{L/4, W/4\}$, $\{L/2, W/2\}$ and $\{3L/4, 3W/4\}$. The excitation port (PORT) is placed at $\{3L/4, W/4\}$ and the pulsed-voltage response is observed at $\boldsymbol{x}^S = \{L/4, 3W/4\}$ (PROBE). Again, the maximum length of the line segments along the circuit's periphery is equal to $ct_\mathrm{w}/20$.

This example demonstrates how the approximation of the ports' shape affects the pulsed-voltage response. Figure 16.7 shows the calculated pulse shapes. As can be seen, all the pulses coincide with each other in the early-time part of the response. Next, it can be seen that the voltage response corresponding to the (rough) 3-point approximation ($P = 3$) starts to significantly deviate from the referential FIT-based signal approximately for $ct > 10\,ct_\mathrm{w}$. The correspondence of this late-time part improves with the increasing number of line segments along the vias' rim. For $P = 12$, a satisfactory correspondence, with respect to the reference, is reached throughout the chosen time window. The remaining discrepancies can be partially attributed to the fact that while in the CIM-based approach we 'measure' the voltage response using the circular probe of (finite) radius $\varrho = 0.50\,[\text{mm}]$, the FIT-based probe spots the response at its center.

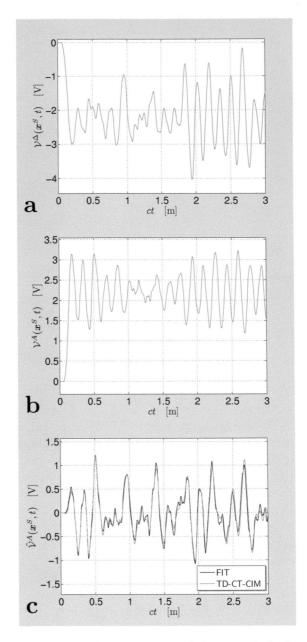

Figure 16.5: The pulsed-voltage responses of the irregularly-shaped circuit observed at $x^S = \{40, 80\}$ [mm]. (a) The impact of the shorting via structures; (b) the voltage response without the via structures; (c) the actual voltage response evaluated using the proposed TD-CT-CIM approach and the referential FIT.

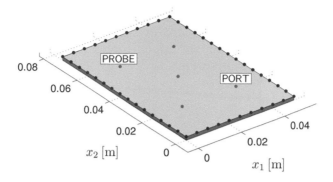

Figure 16.6: The analysed rectangular circuit with shorting via structures.

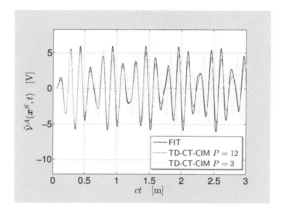

Figure 16.7: The actual voltage response evaluated at $x^S = \{L/4, 3W/4\}$ using the proposed TD-CT-CIM approach for $P = \{3, 12\}$ and the referential FIT.

16.4   Conclusions

An efficient TD-CIM-based computational methodology for analyzing the impact of shorting via structures in parallel-plane structures has been proposed. Namely, it has been demonstrated that the TD compensation theorem for parallel-plane structures is a suitable tool for formulating a system of integral equations of the time-convolution type whose solution yields the induced electric-current pulses on conducting vias. It has been explicitly shown that the system of equations is readily solvable in an updating step-by-step manner with the aid of the standard trapezoidal rule. Subsequently, with the electric-current signals at hand, a straightforward methodology for evaluating the impact of the via structures has been suggested. This methodology employs the trapezoidal approximation, again, and a closed-form deconvolution formula, which makes the procedure computationally efficient and accurate. Finally, sample numerical examples have been discussed and successfully validated using (three-dimensional) FIT.

Appendix A

Integrals of the Logarithmic Function

The two-dimensional Green's function that appears in CIM shows the logarithmic singularity that must be integrated over a line segment of the boundary contour. On account of this, let us consider the following integral

$$I = \int_{\boldsymbol{x} \in \Omega^{AB}} \ln[r(\boldsymbol{x}|\boldsymbol{x}^C)] \mathrm{d}l(\boldsymbol{x}) \tag{A.1}$$

where Ω^{AB} is the line segment determined by points A and B, $r(\boldsymbol{x}|\boldsymbol{x}^C)$ denotes the Eucledian distance between a point on Ω^{AB} and point C lying off the segment (see Fig. A.1). The position of each point is specified by the position vector $\boldsymbol{x}$ (see Sec. 1.2).

Upon introducing the parametrization for $\boldsymbol{x} \in \Omega^{AB}$, that is

$$\boldsymbol{x} = \boldsymbol{x}^A + \lambda(\boldsymbol{x}^B - \boldsymbol{x}^A) \quad \text{for } \lambda \in (0,1) \tag{A.2}$$

we arrive at the one-dimensional integral that can be solved analytically. In this way we end up with

$$I = R^{CP} \Big\{ \tan(\psi^{BC}) \left[\ln(R^{BC}) - 1\right]$$
$$- \tan(\psi^{AC}) \left[\ln(R^{AC}) - 1\right] + \psi^{BC} - \psi^{AC} \Big\} \tag{A.3}$$

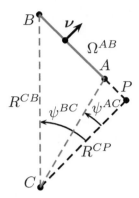

Figure A.1: Integration along the line segment.

where ψ^{AC} and ψ^{BC} are *oriented* angles between the perpendicular line from point C to segment Ω^{AB} and the lines of points C with A and C with B, respectively (see Fig. A.1).

If point C lies on the same line as points A and B then the integral is easily found via the corresponding limiting process. Note that, for such a case, the derivative of the logarithmic function along the direction parallel to the normal vector $\boldsymbol{\nu}$ is zero, that is

$$\partial_\nu \ln[r(\boldsymbol{x}|\boldsymbol{x}^C)] = \boldsymbol{\nu} \cdot (\boldsymbol{x} - \boldsymbol{x}^C)/r^2(\boldsymbol{x}|\boldsymbol{x}^C) = 0 \qquad (\text{A.4})$$

Appendix B

Implementation of TD-CIM

In this section, a demo implementation of TD-CIM in MATLAB® is described. For the sake of simplicity, we shall limit ourselves to the description of a code capable of analyzing an instantaneously-reacting planar circuit (see Chapter 3).

B.1 Geometry of the circuit pattern

At first, the circuit's contour in a plane must be defined. This can be done, for example, by giving two 1D arrays that specify the polygon's vertices in the (x_1, x_2)-plane with respect to the chosen origin. The coordinates of the corners of the circuit shown in Fig. B.1 can be arranged in the following way

```
x = [0 0.10 0.10 0];
y = [0 0 0.15 0.15];
```

along the x_1- and x_2-direction, respectively.

Once the vertices are specified, the circuit's outer rim is divided into N straight-line sections. As a rule of thumb, the maximum length of the sections should be shorter than a tenth of the spatial support of the excitation pulse. Nevertheless, it has been demonstrated that even when this rule is violated, it may well be that the results are accurate enough. The line segments and the dividing points are numbered in

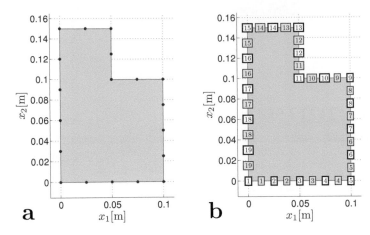

Figure B.1: The sample circuit pattern. (a) The dividing points along the circuit periphery; (b) the numbering of the line segments and the nodal points.

counterclockwise direction as illustrated in Fig. B.1b. The co-ordinates of the dividing nodes of the sample circuit from Fig. B.1a then read

```
X = [0 0.025 0.050 0.075 0.100 0.100 0.100 0.100 0.100 0.075 0.050
     0.050 0.050 0.025 0 0 0 0 0];
Y = [0 0 0 0 0 0.025 0.050 0.075 0.100 0.100 0.100 0.125 0.150 0.150
     0.150 0.120 0.090 0.060 0.030];
```

For possibly multiplying connected circuit patterns, it is convenient to define an auxiliary variable, say IND, that assigns to each line segment the number of its bounding nodes. It is straightforward to find the elements of such a variable automatically for a general circuit geometry. Concerning the example from Fig. B.1, this information can be stored in the following way

```
IND.' =
[1 2 3 4 5 6 7 8 9 10 11 12 13 14 15 16 17 18 19;
 2 3 4 5 6 7 8 9 10 11 12 13 14 15 16 17 18 19 1]
```

Then, for example, IND(19,1) and IND(19,2) return 19 and 1, respectively, which are the numbers of the discretization points bounding segment 19 (see Fig. B.1b). Furthermore, it is also useful to define another variable, say SEG, that assigns to each dividing node the number of its adjacent segments. For the sample circuit given in Fig. B.1 it may have the following form

```
SEG.' =
[19 1 2 3 4 5 6 7 8 9 10 11 12 13 14 15 16 17 18;
  1 2 3 4 5 6 7 8 9 10 11 12 13 14 15 16 17 18 19]
```

For instance, for the first node at the origin we get `SEG(1,1)` = 19 and
`SEG(1,2)` = 1 (cf. Fig. B.1b). With the auxiliary variable at hand, we
may further evaluate the sine and cosine functions of the angle that is
found between the tangent to the n-th line segment (recall the contour
orientation) and the x_1-axis, i.e.

```
for n = 1 : N
 COSW(n) = (X(IND(n,2)) - X(IND(n,1)))/W(n);
 SINW(n) = (Y(IND(n,2)) - Y(IND(n,1)))/W(n);
end
```

where `W(n)` is the length of the n-th segment. Regarding the sample
geometry, this leads to

```
COSW = [1.0 1.0 1.0 1.0 0 0 0 0 -1.0 -1.0 0 0 -1.0 -1.0 0 0 0 0 0];
```

for the array of segment's lengths (cf. Fig. B.1a)

```
W = [0.025 0.025 0.025 0.025 0.025 0.025 0.025 0.025 0.025 0.025
     0.025 0.025 0.025 0.025 0.030 0.030 0.030 0.030 0.030];
```

For example, `find(COSW == -1)` = [9 10 13 14] are then linked to
the numbers of those line segments whose tangent vectors have the
opposite orientation with respect to the x_1-axis (see Fig. B.1b). Arrays
`COSW` and `SINW` will be used to evaluate $\cos(\theta)$ that appears in the
elements of $\boldsymbol{Q}$ array (see Eq. (3.10)).

B.2 Numerical integration

Since the evaluation of $\boldsymbol{Q}$ and $\boldsymbol{F}$ arrays requires computation of line
integrals, it is convenient to specify the position on the k-th line segment
in terms of parameter `lambda` whose values range from 0 to 1, i.e.

```
XS = @(lambda,k) X(IND(k,1)) + lambda*(X(IND(k,2)) - X(IND(k,1)));
YS = @(lambda,k) Y(IND(k,1)) + lambda*(Y(IND(k,2)) - Y(IND(k,1)));
```

where we have used the concept of anonymous functions with two arguments `lambda` and `k`. In the implementation that follows, the former
argument represents the abscissas of the Gauss-Legendre quadrature.
Other numerical integration routines may serve the purpose as well.
For example, we may use the 6-point quadrature [1, p. 921], i.e.

```
WINT.' = [0.0856622462 0.1803807865 0.2339569673 0.2339569673
     0.1803807865 0.0856622462];
LINT.' = [0.0337652429 0.1693953068 0.3806904070 0.6193095930
     0.8306046932 0.9662347571];
```

and define the following 6×6 matrices

```
NI = 6;
LINT1 = repmat(LINT,[1 NI]);
LINT2 = LINT1.';
```

These auxiliary vectors and matrices will be used in the following sections.

B.3 Computation of excitation array F

In this section we describe the implementation of feeding ports formulated in Sec. 2.1.4. For illustrative purposes, we shall start with the simplified model of a circular excitation port whose implementation does not require the handling of the logarithmic singularity.

B.3.1 Singularity-free excitation port

The simplified excitation model evaluates the elements of excitation array F using Eq. (2.49). In this relation, the position of the excitation port is specified by the position vector x^C. The probe must be placed inside the polygon specified by x and y arrays and hence we may take

```
XC = 0.10/4; YC = 0.15/4;
```

In the next step, the electric-current pulse shape is chosen. For example, let us implement the unipolar triangular pulse signature described by Eq. (5.24) (see Fig. 5.2). For such a choice, the relevant Laplace-transform inversion can be carried out analytically, viz.

$$\mathcal{L}^{-1}[s\hat{I}(s)\mathrm{K}_0(sr/c)] = (2A/t_\mathrm{w})[\zeta(r,t) - 2\zeta(r,t-t_\mathrm{w}/2) + \zeta(r,t-t_\mathrm{w})] \tag{B.1}$$

where $\zeta(r,t)$ is defined in Eq. (6.7). After creating the zero excitation array

```
F = zeros(N,NT);
```

and defining the time axis, e.g.

```
cT = linspace(0, 3.0, NT);
```

in which NT is its length, the excitation array's elements may be evaluated for each dividing node in a loop, viz

```
for m = 1 : N
   % ascending +
   R = sqrt((XC - XS(LINT,SEG(m,1))).^2 ...
          + (YC - YS(LINT,SEG(m,1))).^2);
   HLP = repmat(LINT,[1 NT]) .* ETA(R,cT);
   F(m,:) = F(m,:) - (mu/pi)*W(SEG(m,1))*WINT.'*HLP;
   % descending -
   R = sqrt((XC - XS(LINT,SEG(m,2))).^2 ...
          + (YC - YS(LINT,SEG(m,2))).^2);
   HLP = repmat(1-LINT,[1 NT]) .* ETA(R,cT);
   F(m,:) = F(m,:) - (mu/pi)*W(SEG(m,2))*WINT.'*HLP;
end
```

The procedure consists of two similar parts, depending whether we are 'testing' on the 'ascending' or on the 'descending' half of the triangular testing function $T^{[S]}(\boldsymbol{x}^T)$. The latter is represented by LINT and by 1-LINT in auxiliary variable HLP. Next, R is a NI×1 1D array that represents $r(\boldsymbol{x}^C|\boldsymbol{x}^T)$ (viz. Eq. (2.48)), i.e. the array of the Eucledian distances from the circular excitation port to 'testing points' along the relevant line segment. Finally, ETA(R,cT) represents the space-time function given in Eq. (B.1) that returns a NI×NT 2D array evaluated for each distance stored in R and each instant stored in cT. This can be most efficiently done using vectorization.

B.3.2   General excitation port

In the first step, the singular part of Eq. (2.45) as $\boldsymbol{x} \to \boldsymbol{x}^T$ is handled. To this end, we take $\partial \hat{J}_3^B(\boldsymbol{x}^T|\boldsymbol{x}^S, s) = T^{[S]}(\boldsymbol{x}^T)$ and make use of the results introduced in Appendix A. Upon dividing the port's rim $\partial \Omega^P$ into NP elements, i.e. $\partial \Omega^P \simeq \cup_{P=1}^{NP} \triangle \Omega^{[P]}$, we end up with (cf. Eq. (2.45))

$$\left[2s\mu \hat{I}(s)/|\partial \Omega^P|\right] \int_{\boldsymbol{x}^T \in \partial \Omega} T^{[S]}(\boldsymbol{x}^T) \int_{\boldsymbol{x} \in \partial \Omega^P} \hat{G}_\infty[r(\boldsymbol{x}|\boldsymbol{x}^T), s] \mathrm{d}l(\boldsymbol{x}) \mathrm{d}l(\boldsymbol{x}^T)$$

$$\simeq [s\mu \hat{I}(s)/\pi NP] [11/6 - \gamma - \ln(2)/3 + \ln(c/|\triangle \Omega|) - \ln(s)] \qquad (B.2)$$

for $S = P$ or $S = P + 1$ (see Fig. 2.4), where we have assumed $|\triangle \Omega^{[P-1]}| = |\triangle \Omega^{[P]}| = |\triangle \Omega^{[P+1]}| = |\triangle \Omega|$, for simplicity. The remaining integration in Eq. (2.45) is straightforward and does not require any further explanation. After the initialization of F and cT arrays, the procedure starts with a loop over the line segments of the excitation port, i.e.

```
for r = 1 : NP
    SING = ... % evaluated according to (B.2)
    F(IND(r,1),:) = F(IND(r,1),:) - SING;
    F(IND(r,2),:) = F(IND(r,2),:) - SING;
    % to be continued
end
```

where `SING` is an $1 \times$NT array that corresponds to the TD counterpart
of Eq. (B.2). Note, in this respect, that the Laplace-transform inversion can be carried out in closed form for many excitation pulses. An
example from this category is the bell-shaped pulse described in Appendix F, `SING` array of which is calculated using elementary functions
only (see [1, (29.3.62) and (29.3.99) for $k = 2$]). The procedure is then
supplemented by a nested loop over discretization nodes, excluding the
ones corresponding to the r-th segment. An illustrative code may look
as follows

```
for r = 1 : NP
    SING = ... % evaluated according to (B.2)
    F(IND(r,1),:) = F(IND(r,1),:) - SING;
    F(IND(r,2),:) = F(IND(r,2),:) - SING;
    %
    for m = setdiff(1 : N+NP,[r (r+1 - NP*(r==NP))])
        % ascending +
        R = sqrt((XS(LINT1,r) - XS(LINT2,SEG(m,1))).^2 ...
            + (YS(LINT1,r) - YS(LINT2,SEG(m,1))).^2);
        HLP = repmat(LINT2,[1 1 NT]).*ETA(R,cT);
        F(m,:) = F(m,:) - (mu/NP/pi)*W(SEG(m,1))...
            *((reshape(WINT.'*HLP(:,:),[NI NT])).'*WINT).';
        % descending -
        R = sqrt((XS(LINT1,r) - XS(LINT2,SEG(m,2))).^2 ...
            + (YS(LINT1,r) - YS(LINT2,SEG(m,2))).^2);
        HLP = repmat(1-LINT2,[1 1 NT]).*ETA(R,cT);
        F(m,:) = F(m,:) - (mu/NP/pi)*W(SEG(m,2))...
            *((reshape(WINT.'*HLP(:,:),[NI NT])).'*WINT).';
    end
end
```

where we have assumed that $\partial \Omega^P$ is a closed inner contour being oriented in the clockwise direction (see [80, Fig. A4.1]). Here, `ETA(R,cT)`
represents the space-time function $\mathcal{L}^{-1}[s\hat{I}(s)\mathrm{K}_0(sr/c)]$ that returns a
NI$\times$NT 2D array. Concerning the bell-shaped pulse, for example, we
will get

$$\mathcal{L}^{-1}[s\hat{I}(s)\mathrm{K}_0(sr/c)] = (4A/ct_{\mathrm{w}}^2)[\zeta(r,t) - 2\zeta(r,t-t_{\mathrm{w}}/2)$$
$$+ 2\zeta(r,t-3t_{\mathrm{w}}/2) - \zeta(r,t-2t_{\mathrm{w}})] \qquad (\mathrm{B.3})$$

where the relevant $\zeta(r,t)$ is given as

$$\zeta(r,t) = ct\ln\left[ct/r + \left(c^2t^2/r^2 - 1\right)^{1/2}\right]H(t - r/c)$$
$$- r\left(c^2t^2/r^2 - 1\right)^{1/2}H(t - r/c) \qquad\qquad \text{(B.4)}$$

B.4   Computation of system array Q

This section describes a simple implementation of the time-dependent system array Q whose elements will be calculated according to Eq. (3.10). The demo implementation starts by initializing a zero 3D array, i.e.

```
Q = zeros(N,N,NT);
```

The array elements may be evaluated in two nested loops that each run over all the dividing points. As in the previous section, the calculation procedure can be divided into similar parts. Owing to the double integration in Eq. (3.10), we now have four similar blocks. For the sake of brevity, we closely describe only the part that refers to the case when both the 'testing' and 'expansion' take place on the 'ascending' halves of the triangular testing and expansion functions $T^{[S]}(\boldsymbol{x}^T)$ and $T^{[m]}(\boldsymbol{x})$, respectively. Implementation of the remaining combinations is then straightforward. Hence, the procedure may run along the following lines

```
for m = 1 : N
    for n = 1 : N
        % ascending/ascending +/+
        if (m ~= n)
            R = sqrt((XS(LINT1,SEG(n,1)) - XS(LINT2,SEG(m,1))).^2 ...
                + (YS(LINT1,SEG(n,1)) - YS(LINT2,SEG(m,1))).^2);
            COS = ((XS(LINT1,SEG(n,1)) - XS(LINT2,SEG(m,1)))./R) ...
                    * SINW(SEG(n,1)) ...
                    - ((YS(LINT1,SEG(n,1)) - YS(LINT2,SEG(m,1)))./R) ...
                    * COSW(SEG(n,1));
            HLP = repmat(LINT2.*LINT1.*COS,[1 1 NT]).*PSI(R,cT);
            Q(m,n,:) = squeeze(Q(m,n,:)) ...
                    + W(SEG(n,1))*W(SEG(m,1))*(1/pi/cdT) ...
                    *(reshape(WINT.'*HLP(:,:),[NI NT])).'*WINT;
        end
        %
        % + next similar blocks ...
        %
    end
end
```

in which `cdT` is the time step. In order to exclude the 'self-coupling' overlapping terms, it is noted that the array-element calculation is carried out in the relevant `if`-block. In contrast to the previous section, R is now an `NI`×`NI` 2D array that represents $r(\boldsymbol{x}|\boldsymbol{x}^T)$ (viz. Eq. (3.10)), i.e. the array of the Eucledian distances from 'actual-field-expansion points' to 'testing-field points' along the relevant line segments. Variable `COS` calculated on the following line is, again, an `NI`×`NI` 2D array and represents a discrete form $\cos[\theta(\boldsymbol{x}|\boldsymbol{x}^T)]$ appearing in Eq. (3.10). The integrand of the latter equation is consequently stored in the auxiliary variable `HLP`. The latter is composed from the product of (the 'ascending' parts of) the expansion and testing spatial functions `LINT1` and `LINT2`, respectively, of 2D array `COS` and `PSI`. The latter represents the space-time function $\Psi(r,t)$ (viz. Eq. (3.12)) that returns an `NI`×`NI`×`NT` 3D array being evaluated for each distance from R and each time step from `cT`. Again, its evaluation is most efficiently done using vectorization.

B.5 Step-by-step updating procedure

With the excitation array $\boldsymbol{F}$ and system array $\boldsymbol{Q}$ at our disposal, we may proceed with searching for the unknown field array $\boldsymbol{E}$ by solving Eq. (3.8) in a step-by-step manner. To this end, we begin with the square matrix $\boldsymbol{I}$. Starting with its initialization, i.e.

```
I = zeros(N,N);
```

its filling can be done according to Eq. (3.9), viz.

```
for m = 1 : N
    for n = 1 : N
        if (n == m)
            I(m,n) = (W(SEG(m,1)) + W(SEG(m,2)))/3;
        elseif (SEG(m,1) == SEG(n,2))
            I(m,n) = W(SEG(m,1))/6;
        elseif (SEG(m,2) == SEG(n,1))
            I(m,n) = W(SEG(m,2))/6;
        end
    end
end
```

After inverting the matrix on the left-hand side of Eq. (3.8), i.e.

```
M = inv(I - Q(:,:,2));
```

and initializing the `N`×`NT` 2D array allocated for the unknown field distribution

```
E = zeros(N,NT);
E(:,2) = M * F(:,2);
```

we may launch the step-by-step updating procedure. Its implementation may look as follows (cf. Eq. (3.8))

```
for p = 3 : NT
    SUM = zeros(N,1);
    for m = 3 : p
        SUM = SUM + Q(:,:,m) * E(:,p-m+2);
    end
    E(:,p) = M*(F(:,p) + SUM);
end
```

Once the procedure is terminated, E-array contains the desired electric-field (space-time) distribution at the dividing points along the circuit periphery and at time points along the time axis.

B.6 Evaluation of the response in Ω

With the electric-field space-time distribution along $\partial\Omega$ at our disposal, we can evaluate the TD response within circuit's domain Ω. To this end, we make use of Eq. (2.37) and write

$$\hat{V}(\boldsymbol{x}^T, s) = s\mu_0 d\hat{I}(s)\hat{G}_\infty(\boldsymbol{x}^S|\boldsymbol{x}^T, s)$$
$$- \int_{\boldsymbol{x}\in\partial\Omega} \hat{V}(\boldsymbol{x}, s)\partial_\nu\hat{G}_\infty[r(\boldsymbol{x}|\boldsymbol{x}^T), s]\mathrm{d}l(\boldsymbol{x}) \qquad (B.5)$$

for $\boldsymbol{x}^T \in \Omega$, where we have assumed $\hat{V} = -d\hat{E}_3$ and $\hat{J}_3(\boldsymbol{x}, s) = \hat{I}(s)\delta(\boldsymbol{x} - \boldsymbol{x}^S)$, for simplicity. Recall that Eq. (B.5) is not an integral equation but a mere formula, to be evaluated for the (known) voltage distribution found upon executing the step-by-step updating procedure from Sec. B.5. The integration in Eq. (B.5) is approximated as

$$- \int_{\boldsymbol{x}\in\partial\Omega} \hat{V}(\boldsymbol{x}, s)\partial_\nu\hat{G}_\infty[r(\boldsymbol{x}|\boldsymbol{x}^T), s]\mathrm{d}l(\boldsymbol{x})$$
$$\simeq -\sum_{m=1}^{N} \triangle\Omega^{[m]}\, \boldsymbol{\nu}^{[m]} \cdot \boldsymbol{\nabla}\hat{G}_\infty(\boldsymbol{x}^{[m;c]}|\boldsymbol{x}^T, s)\frac{\hat{v}^{[m]}(s) + \hat{v}^{[m+1]}(s)}{2} \qquad (B.6)$$

where the contour integral was replaced by the sum over N straight-line segments, $\boldsymbol{\nu}^{[m]}$ is the unit normal vector of the m-th segment, $\boldsymbol{x}^{[m;c]}$ localizes its center and $\hat{v}^{[m]}(s)$ is the complex-FD counterpart of the TD voltage response pertaining to the m-th dividing node. Next, $\nabla \hat{G}_\infty$ has the meaning of

$$\nabla \hat{G}_\infty(\boldsymbol{x}|\boldsymbol{x}^T, s) = \hat{\boldsymbol{r}}(\boldsymbol{x}|\boldsymbol{x}^T) \partial_r \hat{G}_\infty(\boldsymbol{x}|\boldsymbol{x}^T, s) \tag{B.7}$$

in which $\hat{\boldsymbol{r}}(\boldsymbol{x}|\boldsymbol{x}^T)$ is a unit vector in the direction given by $\boldsymbol{x} - \boldsymbol{x}^T$ and ∂_r denotes the spatial differentiation with respect to r. Owing to the piecewise linear spatial expansion (see Eq. (3.4)), the average of the neighboring nodal values $[\hat{v}^{[m]}(s) + \hat{v}^{[m+1]}(s)]/2$ is the exact result of integration of the voltage distribution over the m-th segment. Making use of the bounded solution of the modified Helmholtz equation in $\mathbb{R}^2$ (see Sec. 6.2 and Appendix J, for instance)

$$\hat{G}_\infty(\boldsymbol{x}|\boldsymbol{x}^T, s) = \mathrm{K}_0[\hat{\gamma}(s)r(\boldsymbol{x}|\boldsymbol{x}^T)]/2\pi \tag{B.8}$$

for $\boldsymbol{x} \neq \boldsymbol{x}^T$ and the expansion (cf. Eq. (3.4))

$$\hat{v}^{[m]}(s) = \sum_{k=1}^{NT} v_{[k]}^{[m]} \hat{T}_{[k]}(s) \tag{B.9}$$

the right-hand side of Eq. (B.6) can be further rewritten as

$$(1/2\pi) \sum_{k=1}^{NT} \hat{T}_{[k]}(s) \sum_{m=1}^{N} \Delta\Omega^{[m]} \boldsymbol{\nu}^{[m]} \cdot \hat{\boldsymbol{r}}(\boldsymbol{x}^{[m;c]}|\boldsymbol{x}^T)$$

$$\hat{\gamma}(s)\mathrm{K}_1[\hat{\gamma}(s)r(\boldsymbol{x}^{[m;c]}|\boldsymbol{x}^T)] \left(\hat{v}_{[k]}^{[m]} + \hat{v}_{[k]}^{[m+1]} \right)/2 \tag{B.10}$$

where we have used [1, (9.6.27)]. The TD counterpart of Eq. (B.10) can be found at once using Eq. (G.3) and we get

$$(1/2\pi c\Delta t) \sum_{k=1}^{NT} \sum_{m=1}^{N} \Delta\Omega^{[m]} \{ \psi[r(\boldsymbol{x}^{[m;c]}|\boldsymbol{x}^T), t - t_{k-1}] - 2\psi[r(\boldsymbol{x}^{[m;c]}|\boldsymbol{x}^T), t - t_k]$$

$$+ \psi[r(\boldsymbol{x}^{[m;c]}|\boldsymbol{x}^T), t - t_{k+1}] \} \boldsymbol{\nu}^{[m]} \cdot \hat{\boldsymbol{r}}(\boldsymbol{x}^{[m;c]}|\boldsymbol{x}^T) \left(\hat{v}_{[k]}^{[m]} + \hat{v}_{[k]}^{[m+1]} \right)/2 \tag{B.11}$$

where (cf. Eqs. (6.19) and (6.24))

$$\psi(r, t) = \mathcal{L}^{-1} \left\{ c\hat{\gamma}(s)\mathrm{K}_1[\hat{\gamma}(s)r]/s^2 \right\} \tag{B.12}$$

which is inverted either numerically via the method introduced in Appendix I or analytically via Eq. (3.13) for an instantaneously-reacting circuit with $\hat{\gamma}(s) = s/c$.

A demo MATLAB® implementation starts with defining the coordinates of the probe, XP and YP, which correspond to the x_1 and x_2 components of $\boldsymbol{x}^T$ in Eq. (B.5), respectively. Subsequently, we fill the arrays of tangential vectors

```
TAU = [COSW.', SINW.', zeros(N,1)];
```

and normal vectors $\boldsymbol{\nu}^{[m]}$

```
NU = cross(TAU,repmat([0 0 1],[N 1]));
```

for all discretization line segments, where variables SINW and COSW have been defined in Sec. B.1. Subsequently, we preallocate two arrays whose elements correspond to distance $r(\boldsymbol{x}^{[m;c]}|\boldsymbol{x}^T)$ and the unit vector $\hat{\boldsymbol{r}}(\boldsymbol{x}^{[m;c]}|\boldsymbol{x}^T)$, i.e.

```
Dr = zeros(N,1); r = zeros(N,3);
```

respectively. Their filling may look as follows

```
for m = 1 : N
    Dr(m) = sqrt((XS(0.5,m) - XP)^2 + (YS(0.5,m) - YP)^2);
    r(m,:) = [XS(0.5,m) - XP, YS(0.5,m) - YP, 0]/Dr(m);
end
```

Voltage values of the integral on the right-hand side of Eq. (B.5) will be stored in a 1×NT 1D array

```
VSCAT = zeros(1,NT);
```

For the sake of clarity, Eq. (B.11) is next evaluated within two for-loops that run along all segments of the approximated boundary and along time axis cT. Such a procedure may look as follows

```
for m = 1 : N
    for k = 1 : NT
        COSF = dot(r(m,:),NU(m,:));
        VA = 0.5*W(m)*(V(IND(m,1),k) + V(IND(m,2),k));
        cTk = cT - (k-1)*cdT; cTk(cTk < 0) = 0;
        psiA = squeeze(PSI(Dr(m),cTk));
        psiB = circshift(psiA,[0 1]); psiB(1) = 0;
        psiC = circshift(psiB,[0 1]); psiC(1) = 0;
        VSCAT = VSCAT + (1/2/pi/cdT)*COSF*(psiA - 2*psiB + psiC)*VA;
    end
end
```

First, variable COSF represents the dot product $\boldsymbol{\nu}^{[m]} \cdot \hat{\boldsymbol{r}}$. Secondly, the average of nodal voltage values pertaining to the m-th segment and k-th time point, multiplied by $\triangle\Omega^{[m]}$, is stored in VA. The shifted time argument $t - t_{k-1}$ is then saved in cTk. The space-time function PSI corresponds to Eq. (B.12) and returns a 1×NT 1D array. This function is called once only for psiA and the values stored in psiB and psiC are found using the circular array shift. Finally, the summation is implemented on the last line that closely resembles Eq. (B.11). Finally note that the space-time voltage distribution along $\partial\Omega$ has been found as

```
V = -d*E;
```

where **E** contains the electric-field space-time distribution along $\partial\Omega$ that has been calculated in Sec. B.5.

The first term on the right-hand side of Eq. (B.5) can be interpreted as the 'primary-wave contribution'. This contribution obviously depends on the choice of the excitation electric-current pulse shape. Concerning the bell-shaped pulse specified in Appendix F, we may write

$$\mathcal{L}^{-1}\left\{s\mu_0 d\hat{I}(s)\hat{G}_\infty(\boldsymbol{x}^S|\boldsymbol{x}^T,s)\right\} = (2A\mu_0 d/\pi t_\mathrm{w})\left\{\phi[r(\boldsymbol{x}^S|\boldsymbol{x}^T),t]\right.$$
$$-2\phi[r(\boldsymbol{x}^S|\boldsymbol{x}^T),t-t_\mathrm{w}/2] + 2\phi[r(\boldsymbol{x}^S|\boldsymbol{x}^T),t-3t_\mathrm{w}/2]$$
$$\left. - \phi[r(\boldsymbol{x}^S|\boldsymbol{x}^T),t-2t_\mathrm{w}]\right\} \tag{B.13}$$

where

$$\phi(r,t) = \mathcal{L}^{-1}\left\{K_0[\hat{\gamma}(s)r]/s^2 t_\mathrm{w}\right\} \tag{B.14}$$

For the Laplace inversion, the method described in Appendix I is applicable again. The special case $\hat{\gamma}(s) = s/c$ can again be handled analytically through

$$\mathcal{L}^{-1}\left\{K_0(sr/c)/s^2 t_\mathrm{w}\right\} = (t/t_\mathrm{w})\ln\left[ct/r + (c^2t^2/r^2-1)^{1/2}\right]H(t-r/c)$$
$$- (r/ct_\mathrm{w})(c^2t^2/r^2-1)^{1/2}H(t-r/c) \tag{B.15}$$

Similarly, for the triangular pulse shape defined in Eq. (5.24) we may write

$$\mathcal{L}^{-1}\left\{s\mu_0 d\hat{I}(s)\hat{G}_\infty(\boldsymbol{x}^S|\boldsymbol{x}^T,s)\right\} = (A\mu_0 d/\pi t_\mathrm{w})\left\{\phi[r(\boldsymbol{x}^S|\boldsymbol{x}^T),t]\right.$$
$$\left. -2\phi[r(\boldsymbol{x}^S|\boldsymbol{x}^T),t-t_\mathrm{w}/2] + \phi[r(\boldsymbol{x}^S|\boldsymbol{x}^T),t-t_\mathrm{w}]\right\} \tag{B.16}$$

where $\phi(r,t)$ must be modified accordingly

$$\phi(r,t) = \mathcal{L}^{-1}\left\{K_0[\hat{\gamma}(s)r]/s\right\} \tag{B.17}$$

with its special case

$$\mathcal{L}^{-1}\left\{K_0(sr/c)/s\right\} = \ln\left[ct/r + (c^2t^2/r^2-1)^{1/2}\right]H(t-r/c) \tag{B.18}$$

being valid for an instantaneously-reacting planar circuit. Other excitation pulse shapes can be incorporated along the same lines. Implementation of Eqs. (B.13) and (B.16) is straightforward and does not require a detailed description.

Appendix C

Implementation of FD-CIM

In this section, demo implementations of the classic FD-CIM in MATLAB® are described. The given description follows Chapter 4, where two CIM-based numerical procedures are derived from the complex-FD reciprocity relation (3.1).

Since the spatial aspects of the problem remain the same as in TD-CIM, the variables defined in Secs. B.1 and B.2 are applicable to the implementation of FD-CIM as well. The difference starts by defining the frequency axis **F** along which the resulting U and H matrices are evaluated. For the frequency range $\{50 \leq f = \omega/2\pi \leq 2000\}$ [MHz], this can be done as follows

```
F = linspace(50,2000,NF)*1e+6;
```

in which NF is the number of frequency points. Alternatively, the logarithmically-spaced frequency axis can be generated using the `logspace` function. The calculations are then carried out in a loop at each frequency point of **F** vector, i.e.

```
for k = 1 : NF
    om = 2*pi*F(k);
    t = sqrt(2/(om*mu0*sigma));
    k1 = om/c; k2 = (k1/2)*(tand + t/d);
    KWN = k1 - 1i*k2;
    % to be continued
end
```

in which **om** stands for the angular frequency ω, **sigma** is electrical conductivity σ of the plates, **t** is their skin depth, **tand** corresponds to $\tan(\delta)$ and finally **KWN** is the complex-valued wave number k (see Eq. (4.17)).

C.1 Computation of U and H matrices

In the first step, we allocate 2D arrays

```
U = zeros(N+NP,N+NP);
H = zeros(N+NP,N+NP);
```

that correspond to U and H, respectively. As has been demonstrated in Chapter 4, computation of the elements of U and H matrices (see Eq. (4.3)) depends on the choice of the testing electric-current surface density $\partial \hat{J}_3^B$ used in the starting reciprocity relation (3.1). Accordingly, this section is divided into two parts.

C.1.1 Point-matching solution

In this section we evaluate the matrix elements according to Eq. (4.5)–(4.8). Having defined the wave number **KWN**, the elements of U and H can be evaluated as follows

```
for k = 1 : NF
    % ...
    % U-matrix and H-matrix
    for m = 1 : N+NP
        for n = 1 : N+NP
            R = sqrt((XS(0.5,m) - XS(0.5,n))^2 ...
                + (YS(0.5,m) - YS(0.5,n))^2);
            COS = ((XS(0.5,n) - XS(0.5,m))*SINW(n) ...
                - (YS(0.5,n) - YS(0.5,m))*COSW(n))/R;
            U(m,n) = -(KWN/2/1i)*COS*besselh(1,2,KWN*R)*W(n);
            H(m,n) = (om*mu0*d/2)*besselh(0,2,KWN*R);
        end
    end
    U(logical(eye(size(U)))) = 1;
    H(logical(eye(size(H)))) = (om*mu0*d/2)*(1-(2*1i/pi) ...
                              *(log(KWN*W/4) - 1 + GAMMA));
    % to be continued
end
```

where the nested loops run over the line segments. In them, R is the distance between the centers of the m-th and n-th segments and XS, YS, W with SINW and COSW have been specified in Secs. B.1 and B.2. The diagonal elements corresponding to the overlapping segments (m = n) are subsequently evaluated in line with Eqs. (4.7) and (4.8), where Euler's constant is represented by GAMMA = 0.577215664901532. Once U and H matrices are known, we can evaluate the impedance matrix Z and the input impedance corresponding to the excitation port, i.e.

```
for k = 1 : NF
    % ...
    % Z-matrix
    Z = U \ H;
    Z11(k) = sum(sum(Z(1:NP,1:NP)))/NP^2;
end
```

where the 'matrix-reduction approach' introduced in [136, Sec. III] is applied to obtain the input impedance Z11 of a port consisting of NP uniformly-excited line segments.

C.1.2   Pulse-matching solution

The use of the rectangular-pulse testing current density results in the matrix elements specified in terms of the double integration taken along the actual and the testing line segments (see Eqs. (4.11) and (4.12)). These integrals are handled using the Gauss-Lengendre quadrature, as described in Sec. B.2. Along these lines, the input impedance can be evaluated as follows

```
for k = 1 : NF
    om = 2*pi*F(k);
    t = sqrt(2/(om*mu0*sigma));
    k1 = om/c; k2 = (k1/2)*(tand + t/d);
    KWN = k1 - 1i*k2;
    % U-matrix and H-matrix
    for m = 1 : N+NP
      for n = 1 : N+NP
        R = sqrt((XS(LINT1,n) - XS(LINT2,m)).^2 ...
            + (YS(LINT1,n) - YS(LINT2,m)).^2);
        COS = ((XS(LINT1,n) - XS(LINT2,m))./R)*SINW(n) ...
            - ((YS(LINT1,n) - YS(LINT2,m))./R)*COSW(n);
        U(m,n) = -(KWN/2/1i)*W(n)*WINT.' ...
                *((COS.*besselh(1,2,KWN*R))*WINT);
        H(m,n) = (om*mu0*d/2) * WINT.'*(besselh(0,2,KWN*R)*WINT);
      end
    end
        U(logical(eye(size(U)))) = 1;
        H(logical(eye(size(H)))) = (om*mu0*d/2)*(1-(2*1i/pi)
```

```
                          *(log(KWN*W/2) - 3/2 + GAMMA));
     % Z-matrix
     Z = U \ H;
     Z11(k) = sum(sum(Z(1:NP,1:NP)))/NP^2;
end
```

Here, R corresponds to $r[\boldsymbol{x}(\lambda)|\boldsymbol{x}^T(\lambda^T)]$ that appears in Eqs. (4.11) and (4.12), COS stands for $\cos\{\theta[\boldsymbol{x}(\lambda)|\boldsymbol{x}^T(\lambda^T)]\}$ and the diagonal terms are calculated via Eqs. (4.15) and (4.16). The final input impedance is calculated at each frequency from 1D-array F using the procedure from [136, Sec. III], again.

Appendix D

Implementation of the Admittance-wall Boundary Condition

The following sample implementation is based on Sec. 6.1, where radiation loss is incorporated via a special case of the admittance-wall boundary condition (6.2). The computational procedure starts by allocating a N×N×NT 3D array that corresponds to the admittance-wall array W, i.e.

```
WW = zeros(N,N,NT);
```

Its elements are subsequently evaluated within two `for`-loops that are running over the dividing points. As in Sec. B.4, the procedure can be divided into four similar blocks that correspond to combinations of the 'ascending' and 'descending' halves of the testing and expansion functions $T^{[S]}(x^T)$ and $T^{[m]}(x)$ (see Eq. (6.5)). An illustrative code may look as follows

```
for m = 1 : N
    for n = 1 : N
        % ascending/ascending +/+
        R = sqrt((XS(LINT1,SEG(n,1)) - XS(LINT2,SEG(m,1))).^2 ...
                + (YS(LINT1,SEG(n,1)) - YS(LINT2,SEG(m,1))).^2) + D;
        HLP = repmat(LINT2,[1 1 NT]).*repmat(LINT1,[1 1 NT]) ...
            .*THETA(R,cT);
        WW(m,n,:) = squeeze(WW(m,n,:)) ...
                + W(SEG(n,1))*W(SEG(m,1))*(nuR/pi/cdT) ...
                *(reshape(WINT.'*HLP(:,:),[NI NT])).'*WINT;
        %
        % + next similar blocks ...
        %
    end
end
```

Here, R is an NI×NI 2D array that contains distances between points on 'the +/+ line segments' of the n-th and m-th nodes. Variables LINT1, LINT2 and WINT with anonymous functions XS, YS have been defined in Sec. B.2. The singularity issue for overlapping line segments is here avoided by the spatial shift D that corresponds to δ shown in Fig. 6.1. Its value is ussually a fraction of the excitation-pulse spatial width ct_w. The (spatial) 3D array R together with the time axis cT are the arguments of the space-time function THETA(R,cT) whose definition is given in Eqs. (6.6) and (6.7). The relative admittance η_0/η that appears in Eq. (6.5) is represented by variable nuR. The value of η_0 is typically a small fraction of the vacuum admittance $(\epsilon_0/\mu_0)^{1/2}$ and $\eta = (\epsilon/\mu_0)^{1/2}$ corresponds to the parameters of the dielectric slab. Once the admittance-wall array W is filled, the launching of the step-by-step updating procedure (viz. Sec. B.5) is preceded by the appropriate modification of system matrix Q, viz

```
Q = Q - WW;
```

thereby accounting for the effect of the radiating boundary (cf. Eqs. (3.8) and (6.4)).

Finally note that the change to the capacitive admittance wall described by Eqs. (6.10) and (6.11) amounts to modifying the space-time function THETA(R,cT) according to Eq. (6.11) and the admittance-wall array, viz

```
%
    WW(m,n,:) = squeeze(WW(m,n,:)) ...
            + W(SEG(n,1))*W(SEG(m,1))*(CR/pi/cdT^2) ...
            *(reshape(WINT.'*HLP(:,:),[NI NT])).'*WINT;
%
```

where variable `CR` corresponds to C_0/ϵ (see Eq. (6.10)). Following the same lines, we may also implement the inductive admittance-wall defined by Eqs. (6.8) and (6.9).

Appendix E

Implementation of Lumped-element Arrays

The following sample implementation is based on Chapter 7, where the incorporation of linear lumped elements into TD-CIM is introduced. The computational procedure starts by allocating an $(N+NP+NC)\times(N+NP+NC)\times NT$ 3D array that corresponds to the lumped-elements array L, i.e.

```
L = zeros(N+NP+NC,N+NP+NC,NT);
```

where NC denotes the number of line segments approximating $\partial\Omega^C$. Recall that NP is the number of line segments along the discretized excitation-port boundary (see Sec. B.3.2) and N refers to the number of line segments along the outer circuit's rim (see Sec. B.1). For the sake of clarity, we shall provide illustrative codes that handle a single lumped element only. An extension to multiple elements and/or their combinations is straightforward.

E.1 Inclusion of a resistor

The lumped-element array concerning a resistor is evaluated according to Eq. (7.10). At first we define variable ADMr that corresponds to Gd/η. Taking into account the uniform electric-current distribution along the port's rim $\partial\Omega^C$, we may write

```
ADMr = (GA/eta)*d/NC;
```

in which **eta** represents $\eta = (\epsilon/\mu)^{1/2}$, **d** is the circuit's thickness and **GA** is element's conductance (see Eq. (7.9)). The procedure is carried out within a **for**-loop over all the (testing) dividing points, i.e.

```
for m = 1 : N+NP+NC
    % ascending +
    R = sqrt((XC - XS(LINT,SEG(m,1))).^2 ...
          + (YC - YS(LINT,SEG(m,1))).^2);
    HLP = repmat(LINT,[1 NT]).*THETA(R,cT);
    HLP_A = W(SEG(m,1))*(ADMr/pi/cdT)*WINT.'*HLP;
    % descending -
    R = sqrt((XC - XS(LINT,SEG(m,2))).^2 ...
          + (YC - YS(LINT,SEG(m,2))).^2);
    HLP = repmat(1-LINT,[1 NT]).*THETA(R,cT);
    HLP_D = W(SEG(m,2))*(ADMr/pi/cdT)*WINT.'*HLP;
    %
    L(m,ND:NU,:) = squeeze(L(m,ND:NU,:)) ...
                 + repmat(HLP_A + HLP_D,[NC 1]);
end
```

Here, **NI**×1 is a 1D array that represents $r(\boldsymbol{x}^C|\boldsymbol{x}^T)$ (viz. Eq. (7.10)), i.e. the array of the Eucledian distances from the circular lumped-element port to testing points' along the relevant line segment. Variables **LINT** and **WINT** with anonymous functions **XS**, **YS** have been defined in Sec. B.2. As in Appendix D, **THETA(R,cT)** is the space-time function whose definition is given in Eqs. (6.6) and (6.7). Finally, integers **ND** and **NU** refers to the indices of the dividing points on $\partial\Omega^C$. Obviously, **NU**−**ND**=**NC**−1.

E.2   Inclusion of an inductor

The lumped-element array concerning an inductor is evaluated according to Eq. (7.12). At first we define variable **INDr** that corresponds to $\mu d/L$. Again, taking into account the uniform electric-current distribution along the port's rim $\partial\Omega^C$, we may write

```
INDr = (mu/LA)*d/NC;
```

in which **mu** is the magnetic permeability, **d** is the circuit's thickness and **LA** is element's inductance (see Eq. (7.11)). The procedure is, again, carried out within the **for**-loop given in the previous section with **ADMr** replaced with **INDr**. In addition, the space-time function **THETA(R,cT)** must be modified accordingly. This amounts to replacing Eq. (6.6) by Eq. (6.9) and accounting for $c\triangle t$ factor.

E.3 Inclusion of a capacitor

The lumped-element array concerning a capacitor is evaluated according to Eq. (7.14). At first we define variable `CAPr` that corresponds to Cd/ϵ. Again, taking into account the uniform electric-current distribution along the port's rim $\partial\Omega^C$, we may write

```
CAPr = (CA/eps)*d/NC;
```

in which `eps` is the electric permittivity, `d` is the circuit's thickness and `CA` is element's capacitance (see Eq. (7.13)). The procedure is, again, carried out within the `for`-loop given in Sec. E.1 with `ADMr` replaced with `CAPr`. In addition, the space-time function `THETA(R,cT)` must be modified accordingly. This amounts to replacing Eq. (6.6) by Eq. (6.11) and accounting for $c\triangle t$ factor.

E.4 Modification of the system array Q

In the final step, once the lumped-element array L is filled, the system matrix Q is before launching the step-by-step updating procedure (viz. Sec. B.5) modified accordingly

```
Q = Q - L;
```

thereby accounting for the effect of a lumped element (cf. Eqs. (3.8) and (7.8)).

Appendix F

The Bell-shaped Pulse

In this section we derive the expression for the excitation pulse signature that is frequently used throughout the book. In general, a *unipolar* pulse can defined by its amplitude A, pulse time width t_w, pulse rise time t_r and if applicable, by its pulse fall time t_f. The pulse time width of the pulse $\mathcal{I}(t)$ is defined as

$$t_w = \int_{t=0}^{\infty} \mathcal{I}(t)\mathrm{d}t/A \qquad \text{(F.1)}$$

and the pulse rise time is defined as the time instant where the pulse reaches its maximum $\mathcal{I}(t_r) = A$. A detailed description of practical waveforms is given by Quak [96].

The *bell-shaped source signature* can be viewed as to be generated by the time convolution of the rectangular function and the triangular function. The basic building block is the rectangular function of the finite duration $\triangle t > 0$ that is defined as

$$\mathrm{R}(t, \triangle t) = \mathrm{H}(t) - \mathrm{H}(t - \triangle t) \qquad \text{(F.2)}$$

where $\mathrm{H}(t)$ is the Heaviside function defined as

$$\mathrm{H}(t) = \begin{cases} 0 & \text{if } t < 0 \\ 1/2 & \text{if } t = 0 \\ 1 & \text{if } t > 0 \end{cases} \qquad \text{(F.3)}$$

Consequently, the triangular function of the duration $\triangle t$ can be found from the time convolution of two rectangular functions, i.e.

$$\mathrm{T}(t, \triangle t) = \mathrm{R}(t, \triangle t/2) * \mathrm{R}(t, \triangle t/2) \qquad \text{(F.4)}$$

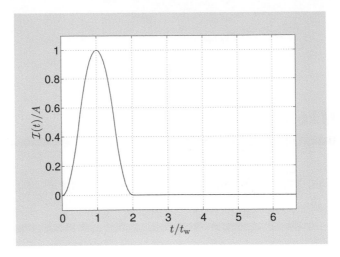

Figure F.1: Time signature of the bell-shaped pulse.

and finally, the bell-shaped function $B(t, \triangle t)$ follows from the time convolution of the triangular function with the rectangular function, i.e.

$$
\begin{aligned}
B(t, \triangle t) &= T(t, \triangle t/2) * R(t, \triangle t/2) \\
&= R(t, \triangle t/4) * R(t, \triangle t/4) * R(t, \triangle t/2) \quad \text{(F.5)}
\end{aligned}
$$

In this manner, we may arrive at the electric current bell-shaped waveform

$$
\begin{aligned}
\mathcal{I}(t) = A \Bigg[&2 \left(\frac{t}{t_w} \right)^2 H(t) - 4 \left(\frac{t}{t_w} - \frac{1}{2} \right)^2 H \left(\frac{t}{t_w} - \frac{1}{2} \right) \\
&+ 4 \left(\frac{t}{t_w} - \frac{3}{2} \right)^2 H \left(\frac{t}{t_w} - \frac{3}{2} \right) - 2 \left(\frac{t}{t_w} - 2 \right)^2 H \left(\frac{t}{t_w} - 2 \right) \Bigg] \quad \text{(F.6)}
\end{aligned}
$$

that is for illustration shown in Fig. F.1. Evidently, the bell-shaped pulse is continuously differentiable; its rise time is equal to the pulse time width and has a finite support, which is suitable for modeling causal wave phenomena.

Appendix G

Expansion Functions

In this section we describe expansion functions $\{T_{[k]}, B_{[k]}, Q_{[k]}\}(t)$ used in the main text for the temporal expansion of the unknown field quantity. To this end, let us consider a temporal function $U(t)$ that is approximated by a set of expansion functions $F_{[k]}(t)$ according to

$$U(t) \simeq \sum_{k=1}^{NT} c_{[k]} F_{[k]}(t) \tag{G.1}$$

where $c_{[k]}$ are coefficients and $F_{[k]}(t)$ can stand for $\{T_{[k]}, B_{[k]}, Q_{[k]}\}(t)$. All these expansion functions have a finite support extending over two time steps $\text{supp}(F_{[k]}) = 2\triangle t$, $\triangle t > 0$, they have value one at one of the discrete time points $F_{[k]}(k\triangle t) = 1$ and zero in the neighboring points $F_{[k]}[(k \pm 1)\triangle t] = 0$ and differ in their differentiability class.

G.1   Linear expansion functions

The linear expansion function (also called as triangular or hat function) can be described as

$$T_{[k]}(t) = \big[(t - t_{k-1})\text{H}(t - t_{k-1}) - 2(t - t_k)\text{H}(t - t_k)$$
$$+ (t - t_{k+1})\text{H}(t - t_{k+1})\big]/\triangle t \tag{G.2}$$

where $\text{H}(t)$ is the Heaviside function (see Eq. (F.3)), $t_k = k\triangle t$ and $k = \{1, ..., NT\}$. The sequence of shifted linear expansion functions and their sum is shown in Fig. G.1. Upon taking the Laplace transform (see Eq. (1.2)) of Eq. (G.2) we arrive at

$$\hat{T}_{[k]}(s) = \big[\exp(-st_{k-1}) - 2\exp(-st_k) + \exp(-st_{k+1})\big]/s^2\triangle t \quad (G.3)$$

A function expanded using the set of $\{T_{[k]}(t), k = 1, 2, ...\}$ is continuous but its first derivative shows jumps at a finite number of time instants, i.e. the function is of class C^0.

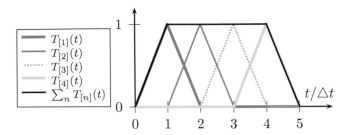

Figure G.1: Linear expansion functions.

G.2 Quadratic expansion functions

The quadratic expansion function can be described as

$$B_{[k]}(t) = 2\big[(t - t_{k-1})^2 H(t - t_{k-1}) - 2(t - t_{k-1/2})^2 H(t - t_{k-1/2})$$
$$+ 2(t - t_{k+1/2})^2 H(t - t_{k+1/2}) - (t - t_{k+1})^2 H(t - t_{k+1})\big]/(\triangle t)^2$$
$$(G.4)$$

where $H(t)$ is the Heaviside function (see Eq. F.3), $t_k = k\triangle t$ and $k = \{1, ..., NT\}$. The sequence of shifted quadratic expansion functions and their sum are shown in Fig. G.2. Upon taking the Laplace transform (see Eq. (1.2)) of Eq. (G.4) we arrive at

$$\hat{B}_{[k]}(s) = 4\big[\exp(-st_{k-1}) - 2\exp(-st_{k-1/2})$$
$$+ 2\exp(-st_{k+1/2}) - \exp(-st_{k+1})\big]/s^3(\triangle t)^2$$
$$(G.5)$$

A function expanded using the set of $\{B_{[k]}(t), k = 1, 2, ...\}$ is continuously differentiable, i.e. the function is of class C^1.

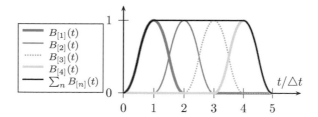

Figure G.2: Quadratic expansion functions.

G.3 Cubic expansion functions

The cubic expansion function can be described as

$$
\begin{aligned}
Q_{[k]}(t) = 16\big[&(t - t_{k-1})^3\mathrm{H}(t - t_{k-1}) - 2(t - t_{k-3/4})^3\mathrm{H}(t - t_{k-3/4}) \\
&+2(t - t_{k-1/4})^3\mathrm{H}(t - t_{k-1/4}) - 2(t - t_k)^3\mathrm{H}(t - t_k) \\
&+2(t - t_{k+1/4})^3\mathrm{H}(t - t_{k+1/4}) - 2(t - t_{k+3/4})^3\mathrm{H}(t - t_{k+3/4}) \\
&+(t - t_{k+1})^3\mathrm{H}(t - t_{k+1})\big]/3(\triangle t)^3
\end{aligned}
\tag{G.6}
$$

where $\mathrm{H}(t)$ is the Heaviside function (see Eq. (F.3)), $t_k = k\triangle t$ and $k = \{1, ..., NT\}$. The sequence of shifted cubic expansion functions and their sum are shown in Fig. G.3.

Upon taking the Laplace transform (see Eq. (1.2)) of Eq. (G.6) we arrive at

$$
\begin{aligned}
\hat{Q}_{[k]}(s) = 32\big[&\exp(-st_{k-1}) - 2\exp(-st_{k-3/4}) + 2\exp(-st_{k-1/4}) \\
&- 2\exp(-st_k) + 2\exp(-st_{k+1/4}) - 2\exp(-st_{k+3/4}) \\
&+ \exp(-st_{k+1})\big]/s^4(\triangle t)^3
\end{aligned}
\tag{G.7}
$$

A function expanded using the set of $\{Q_{[k]}(t), k = 1, 2, ...\}$ is continuous including its second derivative, i.e. the function is of class C^2.

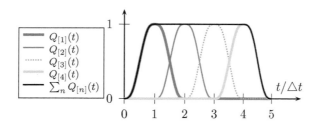

Figure G.3: Cubic expansion functions.

Appendix H

Green's Function of the Dissipative Scalar 2D Wave Equation

In this section we derive the causal solution $G_\infty = G_\infty(\boldsymbol{x}, t)$ of the two-dimensional dissipative wave equation

$$(\partial_1^2 + \partial_2^2)G_\infty - c^{-2}(\partial_t^2 + 2\alpha\partial_t)G_\infty = -\delta(\boldsymbol{x}, t) \qquad \text{(H.1)}$$

with the zero-value initial conditions $G_\infty(\boldsymbol{x}, 0) = 0$ and $\partial_t G_\infty(\boldsymbol{x}, 0) = 0$ for all $\boldsymbol{x} \in \mathbb{R}^2$. Here, c is the wave speed (a positive and real-valued constant) and α is a real non-negative real constant accounting for conductive (diffusive) losses.

The dissipative wave equation (H.1) is solved with the aid of a modification of the Cagniard-DeHoop method [20]. To this end, the one-sided Laplace transformation (1.2) is combined with the dissipative-wave slowness field representation either along x_1- or x_2-direction. For the former direction we take

$$\hat{G}_\infty(x_1, x_2, s) = \frac{\hat{L}(s)}{2\pi\mathrm{i}} \int_{p=-\mathrm{i}\infty}^{\mathrm{i}\infty} \exp\left[-\hat{L}(s)px_1\right] \tilde{G}_\infty(p, x_2, s)\mathrm{d}p \qquad \text{(H.2)}$$

where $\hat{L}(s) = [s(s + 2\alpha)]^{1/2}$ and ∂_1 is transformed to $\tilde{\partial}_1 = -\hat{L}(s)p$. The square-root expression shows two algebraic branch points at $s = \{-2\alpha, 0\}$ in the complex s-plane. The corresponding branch cuts are chosen such that $\mathrm{Re}(s^{1/2}) > 0$ and $\mathrm{Re}[(s + 2\alpha)^{1/2}] > 0$ for all $s \in$

$\mathbb{C}$, which introduces two overlapping branch cuts running along the negative real axis $\{s \in \mathbb{C}; -\infty < \mathrm{Re}(s) < 0, \mathrm{Im}(s) = 0\}$ and $\{s \in \mathbb{C}; -\infty < \mathrm{Re}(s) < -2\alpha, \mathrm{Im}(s) = 0\}$, respectively. The choice of the branch cuts then implies an asymptotic expansion $\hat{L}(s) = s + \mathcal{O}(1)$ as $|s| \to \infty$.

Owing to the slowness representation (H.2) the transform-domain dissipative wave equation transforms to

$$\partial_2^2 \tilde{G}_\infty - \hat{L}^2(s)\gamma^2(p)\tilde{G}_\infty = -\delta(x_2) \tag{H.3}$$

where $\gamma(p)$ is the propagation coefficient

$$\gamma(p) = (1/c^2 - p^2)^{1/2} \quad \text{with} \quad \mathrm{Re}(.)^{1/2} > 0 \tag{H.4}$$

The bounded transform-domain solution of (H.3) is then substituted in (H.2), which finally yields

$$\hat{G}_\infty(x_1, x_2, s) = \frac{1}{2\pi i} \int_{p=-i\infty}^{i\infty} \exp\left\{-\hat{L}(s)\left[px_1 + \gamma(p)|x_2|\right]\right\} \frac{dp}{2\gamma(p)} \tag{H.5}$$

The next few steps follow the classical procedure of the Cagniard-DeHoop method. The original contour in the complex p-plane is deformed into the Cagniard-DeHoop contour defined as

$$px_1 + \gamma(p)|x_2| = \tau \tag{H.6}$$

where $\{\tau \in \mathbb{R}; r/c \leq \tau < \infty\}$ and $r = (x_1^2 + x_2^2)^{1/2} > 0$. The deformation is admissible by virtue of Cauchy's theorem and Jordan's lemma. After the mapping of the variable of integration from the complex p-plane to the real τ-axis we find

$$\hat{G}_\infty(x_1, x_2, s) = \frac{1}{2\pi} \int_{\tau=r/c}^{\infty} \exp\left[-\hat{L}(s)\tau\right] \frac{d\tau}{(\tau^2 - r^2/c^2)^{1/2}} \tag{H.7}$$

Note that Eq. (H.7) formally resembles the result of the standard Cagniard-DeHoop method [20, Eq. (2.18)] for the loss-free wave motion when $\hat{L}(s)$ is equal to s. Now, however, we use [1, (29.3.96)] and Eq. (H.7) is transformed into TD

$$G_\infty(x_1, x_2, t) = (\alpha/2\pi) \int_{\tau=r/c}^{t} \tau \mathrm{I}_1\left[\alpha(t^2 - \tau^2)^{1/2}\right] \exp(-\alpha t)$$
$$(t^2 - \tau^2)^{-1/2}(\tau^2 - r^2/c^2)^{-1/2} d\tau$$
$$+ (1/2\pi)(t^2 - r^2/c^2)^{-1/2} \mathrm{H}(t - r/c) \exp(-\alpha t) \tag{H.8}$$

where $I_1(.)$ is the modified Bessel function of the first kind and the first order. Note that the integral in Eq. (H.8) shows the inverse square-root singularities in both upper and lower limits of the integration that can be extracted through the following substitution

$$\tau^2 = (r/c)^2 \cos^2(\psi) + t^2 \sin^2(\psi) \quad \text{for} \quad 0 \leq \psi \leq \pi/2 \tag{H.9}$$

With the aid of Eq. (H.9) we then arrive at

$$G_\infty(x_1, x_2, t) = (\alpha/2\pi) \exp(-\alpha t) H(t - r/c)$$
$$\int_{\psi=0}^{\pi/2} I_1 \left[\alpha(t^2 - r^2/c^2)^{1/2} \cos(\psi) \right] d\psi$$
$$+ (1/2\pi)(t^2 - r^2/c^2)^{-1/2} H(t - r/c) \exp(-\alpha t) \tag{H.10}$$

The integral in (H.10) can be carried out in closed form, viz

$$\int_{\psi=0}^{\pi/2} I_1 \left[\beta \cos(\psi) \right] d\psi = 2 \sinh^2(\beta/2)/\beta \tag{H.11}$$

with $\{\beta \in \mathbb{R}; \beta > 0\}$, which finally leads to

$$G_\infty(x_1, x_2, t) = (1/2\pi)(t^2 - r^2/c^2)^{-1/2} H(t - r/c)$$
$$\left\{ 1 + 2 \sinh^2[(\alpha/2)(t^2 - r^2/c^2)^{1/2}] \right\} \exp(-\alpha t) \tag{H.12}$$

An alternative way to obtain (H.12) is to start with the corresponding three-dimensional solution [24, Sec. 26.5] and apply Hadamard's method of descent [17, III - §4.4].

Appendix I

Numerical Inversion of the Laplace Transformation

In order to account for the relaxation behavior of planar circuits, a numerical inversion of the Laplace transformation is employed in the main text. This inversion is based on the Bromwich integral

$$f(t) = \frac{1}{2\pi i} \int_{s \in \mathcal{B}} \exp(st) \hat{F}(s) \mathrm{d}s \tag{I.1}$$

for $t > 0$, where $\mathcal{B}$ is the Bromwich integration contour that runs to the right of all singularities of $\hat{F}(s)$. The Bromwich integration contour may be deformed into an equivalent contour $\Gamma \cup \Gamma^*$ (* denotes complex conjugate), provided that $|\hat{F}(s)| \to 0$ in $\mathrm{Re}(s) < 0$ as $|s| \to \infty$ and a new integration contour encloses all singularities of $\hat{F}(s)$ in view of Jordan's lemma and Cauchy's theorem, respectively. A promising candidate is the hyperbolic integration contour defined according to

$$\Gamma = \{s(v) = \sigma_0 - \sigma \cosh(v) + i\nu \sinh(v)\} \tag{I.2}$$

for $\{v \in \mathbb{R}; 0 \le v \le v_\infty\}$, where $\sigma_0 \in \mathbb{R}$, $\{\sigma, \nu \in \mathbb{R}; \sigma > 0, \nu > 0\}$, $v_\infty = \mathrm{acosh}[(\sigma_0 + \sigma_\infty)/\sigma]$ with $\mathrm{Re}[s(v_\infty)] = -\sigma_\infty$. Upon combining the upper and lower halves of the integration contour we arrive at

$$f(t) = \frac{1}{\pi} \mathrm{Im} \int_{v=0}^{v_\infty} \exp[s(v)t] \hat{F}[s(v)] \frac{\mathrm{d}s}{\mathrm{d}v} \mathrm{d}v \tag{I.3}$$

where the relevant Jacobian reads

$$\mathrm{d}s/\mathrm{d}v = -\sigma \sinh(v) + i\nu \cosh(v) \tag{I.4}$$

An efficient way to solve (I.3) is the trapezoidal rule that leads to the following approximation

$$f(t) \simeq \frac{1}{\pi} \frac{v_\infty}{2N} \mathrm{Im} \left\{ \hat{G}[s(0)] + 2 \sum_{k=1}^{N-1} \hat{G}[s(v_k)] + \hat{G}[s(v_\infty)] \right\} \qquad (I.5)$$

where

$$\hat{G}[s(v)] = \exp[s(v)t]\hat{F}[s(v)][\mathrm{d}s(v)/\mathrm{d}v] \qquad (I.6)$$

and $v_k = kv_\infty/N$ for $k = \{0, ..., N\}$.

The applicability of the inversion algorithm can be demonstrated on a relatively simple example that admits the closed-form original $f(t)$. To this end, we shall apply the method to the following s-domain function

$$\hat{F}(s) = \frac{\alpha}{s} \frac{\exp\{-q[s(s+\alpha)]^{1/2}\}}{[s(s+\alpha)]^{1/2}} \qquad (I.7)$$

with $\{\alpha, q \in \mathbb{R}; \alpha, q > 0\}$. Making use of [1, Eq. (29.3.91)], the relevant TD counterpart is written as

$$f(t) = \alpha \int_{\tau=q}^{t} I_0[\alpha(\tau^2 - q^2)^{1/2}/2] \exp(-\alpha\tau/2)\mathrm{d}\tau \qquad (I.8)$$

for $t > \tau$, where $I_0(.)$ is the modified Bessel function of the first kind and the zero order. Similarly to the functions encountered in Sec. 6.2, $\hat{F}(s)$ has two branch points at $s = \{-\alpha, 0\}$ along with a simple pole singularity at $s = 0$. Again, we choose to take the (finite) branch cut that connects the two branch points (see Fig. 6.2). For a demo implementation in MATLAB® we take

```
alpha = 2.0e+6;
q = 1.0e-6;
```

for example, which stands for $\alpha = 2.0 \cdot 10^6 \, [\mathrm{s}^{-1}]$ and $q = 1.0 \cdot 10^{-6} \, [\mathrm{s}]$. The inversion of Eq. (I.7) will be evaluated along the time axis that extends from q to $10q$. Hence, we define

```
time = linspace(q, 10*q, NT);
```

and we may take NT = 1.0e+3, for example. In the next step, the parameters of the hyperbolic integration contour are defined as

```
sigma = 1.0*alpha;
nu = 1.50*alpha;
sigma0 = 1.05*alpha;
```

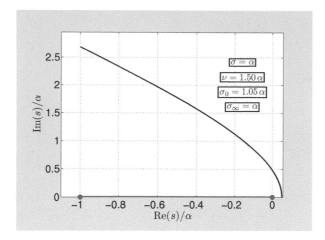

Figure I.1: The hyperbolic integration contour Γ in the second quadrant of the complex s-plane.

which corresponds to $\sigma = \alpha$, $\nu = 1.50\,\alpha$ and $\sigma_0 = 1.05\,\alpha$, respectively. Note that $\sigma_0 > \sigma$ in order to avoid crossing the branch cut. The parametrization of the contour via v is terminated at $v = v_\infty$ that is found from σ_∞, i.e.

```
sigma00 = 1.0*alpha;
v00 = acosh((sigma0 + sigma00)/sigma);
```

where `sigma00` stands for σ_∞ and v_∞ is represented by `v00`. The corresponding hyperbolic contour in the second quadrant of the (normalized) complex s-plane is shown in Fig. I.1. The parametrized integration contour with the Jacobian (see Eqs. (I.2) and (I.4)) can be implemented via MATLAB® function handles, i.e.

```
sH = @(v) sigma0 - sigma*cosh(v) + 1i*nu*sinh(v);
JAC = @(v) - sigma*sinh(v) + 1i*nu*cosh(v);
```

respectively. Starting with the initialization

```
N = 1.0e+3;
FN = zeros(1, NT);
FS = @(s) alpha*exp(-q*sqrt(s.*(s + alpha))) ...
        ./sqrt(s.*(s + alpha))./s;
```

the trapezoidal-rule formula (I.5) can be rewritten along the following lines

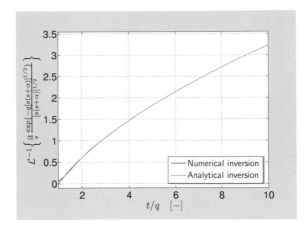

Figure I.2: The TD counterpart of Eq. (I.7) evaluated using the method based on the deformation of the Bromwich contour into the hyperbolic path ('Numerical inversion') and using Eq. (I.8) ('Analytical inversion').

```
for k = 0 : N
    vk = k*v00/N;
    svk = sH(vk);
    sdvk = JAC(vk) - 0.5*JAC(vk)*(k == 0 || k == N);
    H = FS(svk); H = repmat(H, [1 NT]);
    E = exp(svk*time);
    FN = FN + imag(H.*E*sdvk);
end
FN = (1/pi)*(v00/N)*FN;
```

Here, the desired result $f(t)$ is stored in a 1×NT 1D array called FN. The complex-FD function $\hat{F}(s)$ is represented by function FS. In our illustrative implementation, the latter is equal to Eq. (I.7). For-loop counter k directly corresponds to k in the summation formula (I.5). Here, variable vk represents $v_k = k v_\infty / N$ and svk is the corresponding complex-frequency point lying on the integration contour Γ parametrized via function sH. Variable sdvk represents the parametrized Jacobian $\mathrm{d}s(v)/\mathrm{d}v$ for $v = v_k$. This variable further accounts for the halving of the end terms in the trapezoidal-rule formula. The evaluation of the complex-FD function itself at a complex-frequency point $s(v_k)$ is executed by FS(svk). Its result is stored in variable H. In accordance with Eq. (I.6), variables H and E contain $\hat{F}[s(v_k)]$ and $\exp[s(v_k)t]$, respectively. Finally, the summation and the constant multiplication is executed in line with Eq. (I.5).

The outcomes of the presented numerical inversion are compared with the results obtained via formula (I.8) in Fig. I.2. The latter has been evaluated using the adaptive Gauss-Kronrod quadrature as implemented in MATLAB®. Upon introducing the integrated function IF and a 1×NT 1D array called **FA**

```
IF = @(tau) besseli(0,alpha*sqrt(tau.^2-q^2)/2).*exp(-alpha*tau/2);
FA = zeros(1, NT);
```

the integral can be for each **time(k)** easily evaluated in a **for**-loop, i.e.

```
for k = 1 : NT
    FA(k) = alpha*quadgk(IF,q,time(k));
end
```

Note that the integration can be readily carried out with any prescribed accuracy. The default values of the absolute and relative error tolerances are **1.0e-10** and **1.0e-6**, respectively, for the double precision. The apparent discrepancy at the early-time part of $f(t)$ can be reduced by changing the parameters of the integration contour and/or by increasing the number of integration points **N** in the summation formula. Concerning the s-domain functions appearing in the TD-CIM formulation (see Eq. (6.12)), potential early-time errors can be eliminated by matching the modified Bessel functions with sufficiently smooth temporal expansion functions and excitation pulse shapes.

Appendix J

Green's Function of the Scalar 2D Wave Equation with Relaxation

Consider the following wave equation

$$(\partial_1^2 + \partial_2^2)G_\infty - c^{-2}\partial_t^2(G_\infty + \chi * G_\infty) = -\delta(\boldsymbol{x}, t) \qquad \text{(J.1)}$$

with the zero-value initial conditions $G_\infty(\boldsymbol{x}, 0) = 0$ and $\partial_t G_\infty(\boldsymbol{x}, 0) = 0$ for all $\boldsymbol{x} \in \mathbb{R}^2$. Here, c is the wave speed (a positive and real-valued constant) and $\chi = \chi(t)$ is the (causal) relaxation function, i.e. $\chi(t) = 0$ for all $t < 0$.

The wave equation (J.1) is solved with the aid of a modification of the Cagniard-DeHoop method [20]. To this end, the one-sided Laplace transformation (1.2) is combined with the dispersive-wave slowness field representation either along x_1- or x_2-direction. For the former direction we take

$$\hat{G}_\infty(x_1, x_2, s) = \left[s(1 + \hat{\chi})^{1/2}\right] / 2\pi \mathrm{i}$$

$$\int_{p=-\mathrm{i}\infty}^{\mathrm{i}\infty} \exp\left[-s(1 + \hat{\chi})^{1/2}px_1\right] \tilde{G}_\infty(p, x_2, s)\mathrm{d}p \qquad \text{(J.2)}$$

where $\hat{\chi} = \hat{\chi}(s)$ is real and positive for real and positive values of s and monotonically decreases toward zero as $s \to \infty$. As a consequence of the slowness representation, ∂_1 is transformed to $\tilde{\partial}_1 = -s(1 + \hat{\chi})p$ and the transform-domain wave equation reads

$$\partial_2^2 \tilde{G}_\infty - s^2(1 + \hat{\chi})\gamma^2(p)\tilde{G}_\infty = -\delta(x_2) \qquad (J.3)$$

where $\gamma(p)$ is the propagation coefficient

$$\gamma(p) = (1/c^2 - p^2)^{1/2} \quad \text{with} \quad \text{Re}(.)^{1/2} > 0 \qquad (J.4)$$

The bounded transform-domain solution of (J.3) is then substituted in (J.2), which finally yields

$$\hat{G}_\infty(x_1, x_2, s) = (1/2\pi i)$$
$$\int_{p=-i\infty}^{i\infty} \exp\left\{-s(1 + \hat{\chi})^{1/2}[px_1 + \gamma(p)|x_2|]\right\} dp/2\gamma(p) \qquad (J.5)$$

The next few steps follow the standard procedure of the Cagniard-DeHoop method. The original contour in the complex p-plane is deformed into the Cagniard-DeHoop contour defined as

$$px_1 + \gamma(p)|x_2| = \tau \qquad (J.6)$$

where $\{\tau \in \mathbb{R}; r/c \le \tau < \infty\}$ and $r = (x_1^2 + x_2^2)^{1/2} > 0$. The deformation is admissible by virtue of Cauchy's theorem and Jordan's lemma. After the mapping of the variable of integration from the complex p-plane to the real τ-axis we find

$$\hat{G}_\infty(x_1, x_2, s) = \frac{1}{2\pi} \int_{\tau=r/c}^{\infty} \exp\left[-s(1 + \hat{\chi})^{1/2}\tau\right] \frac{d\tau}{(\tau^2 - r^2/c^2)^{1/2}} \qquad (J.7)$$

Note that up to this point, the described procedure is similar to the one for the dissipative wave equation from Appendix H. Now, however, we assume a general relaxation behavior, for which the exponential kernel does not have a closed form Laplace-transform inversion. For such a case we start with the Bromwich inversion integral and write

$$E(t, \tau) = \frac{1}{2\pi i} \int_{s \in \mathcal{B}} \exp(st) \exp\left[-\hat{L}(s)\tau\right] ds \qquad (J.8)$$

where $\hat{L}(s) = s(1 + \hat{\chi})^{1/2}$ as used in Appendix H and $\mathcal{B}$ is the Bromwich contour. The Bromwich contour is parallel with $\text{Re}(s) = 0$ and is shifted to the right of all singularities in the complex s-plane. For standard relaxation models, $\hat{L}(s)$ shows algebraic branch points due to square roots for which we choose $\text{Re}[(.)^{1/2}] > 0$ for all $s \in \mathbb{C}$. This implies the (overlapping) branch cuts along the negative real axis in the complex s-plane. Then, the following asymptotic expansion holds

$$\hat{L}(s) = \hat{L}_\infty(s) + \mathcal{O}(s^{-1}) \tag{J.9}$$

as $|s| \to \infty$, where $\hat{L}_\infty(s) = s + w$, $\{w \in \mathbb{R}; w \geq 0\}$. In virtue of Jordan's lemma we therefore rewrite (J.8) as

$$
\begin{aligned}
E(t,\tau) &= \frac{1}{2\pi\mathrm{i}} \int_{s \in \mathcal{B}} \exp(st) \left\{ \exp\left[-\hat{L}(s)\tau\right] - \exp\left[-\hat{L}_\infty(s)\tau\right] \right\} \mathrm{d}s \\
&+ \frac{1}{2\pi\mathrm{i}} \int_{s \in \mathcal{B}} \exp(st) \exp\left[-\hat{L}_\infty(s)\tau\right] \mathrm{d}s
\end{aligned}
\tag{J.10}
$$

The second term in Eq. (J.10) represents the (attenuated) Dirac delta distribution

$$\frac{1}{2\pi\mathrm{i}} \int_{s \in \mathcal{B}} \exp(st) \exp\left[-\hat{L}_\infty(s)\tau\right] \mathrm{d}s = \delta(t - \tau) \exp(-w\tau) \tag{J.11}$$

while the first term can be handled numerically along the lines described in Appendix I. In this process, the Bromwich contour is for $t > \tau$ closed to the right by supplementing it with a semi-circle of radius $\Delta \to \infty$ and the resulting contour is, in view of Cauchy's theorem, contracted into the hyperbolic contour $\Gamma \cup \Gamma^*$, provided that all singularities are enclosed. Once the numerical integration is carried out, we arrive at

$$E(t,\tau) = F(t,\tau)\mathrm{H}(t - \tau) + \delta(t - \tau) \exp(-w\tau) \tag{J.12}$$

Upon substituting Eq. (J.12) into the TD counterpart of (J.7) we end up with

$$
\begin{aligned}
G_\infty(x_1, x_2, t) &= (1/2\pi)(t^2 - r^2/c^2)^{-1/2}\mathrm{H}(t - r/c) \exp(-wt) \\
&+ (1/2\pi) \int_{\tau=r/c}^{t} (\tau^2 - r^2/c^2)^{-1/2} F(t,\tau) \mathrm{d}\tau
\end{aligned}
\tag{J.13}
$$

where we have used the sifting property of the Dirac distribution. The first part in Eq. (J.13) is the (attenuated) fundamental solution of the two-dimensional wave equation, while the second one represents the effect of (Boltzmann-type) relaxation. The latter is negligible close to the wavefront as $t \downarrow r/c$, which is a general feature of dispersive phenomena [37]. Finally, note that the dissipative wave equation solved in Appendix H is a special case of Eq. (J.1) for $\chi(t) = 2\alpha\mathrm{H}(t)$.

References

[1] M. Abramowitz and I. A. Stegun. *Handbook of Mathematical Functions.* New York, NY: Dover Publications, 1972.

[2] R. Achar and M. S. Nakhla. Simulation of high-speed interconnects. *Proceedings IEEE*, 89(5):693–727, May 2001.

[3] G. Antonini. A low-frequency accurate cavity model for transient analysis of power-ground structures. *IEEE Transactions on Electromagnetic Compatibility*, 50(1):138–148, February 2008.

[4] C. A. Balanis. *Antenna Theory, 3rd Ed.* Hoboken, NJ: John Wiley & Sons, Inc., 2005.

[5] C. E. Baum. General properties of antennas. Sensor and Simulation Notes - Note 330, Philips Lab., Kirkland AFB, NM, July 1991.

[6] C. E. Baum. General properties of antennas. *IEEE Transactions on Electromagnetic Compatibility*, 44(1):18–24, February 2002.

[7] C. L. Bennett and W. L. Weeks. Transient scattering from conducting cylinders. *IEEE Transactions on Antennas and Propagation*, 18(5):627–633, September 1970.

[8] P. Bernardi and R. Cicchetti. Response of a planar microstrip line excited by an external electromagnetic field. *IEEE Transactions on Electromagnetic Compatibility*, 32(2):98–105, May 1990.

[9] M. Bonnet, G. Maier, and C. Polizzotto. Symmetric Galerkin boundary element method. *Applied Mechanics Reviews*, 51:669–704, 1998.

[10] M. Bozzi, A. Georgiadis, and K. Wu. Review of substrate-integrated waveguide circuits and antennas. *IET Microwaves, Antennas & Propagation*, 5(8):909–920, 2011.

[11] M. Bozzi, L. Perregrini, and K. Wu. Modeling of conductor, dielectric, and radiation losses in substrate integrated waveguide by the boundary integral-resonant mode expansion method. *IEEE Transactions on Microwave Theory and Techniques*, 56(12):3153–3161, December 2008.

[12] A. C. Cangellaris and R. Lee. Finite element analysis of electromagnetic scattering from inhomogeneous cylinders at oblique incidence. *IEEE Transactions on Antennas and Propagation*, 39(5):645–650, May 1991.

[13] Y. Cassivi, L. Perregrini, P. Arcioni, M. Bressan, K. Wu, and G. Conciauro. Dispersion characteristics of substrate integrated rectangular waveguide. *IEEE Microwave and Wireless Components Letters*, 12(9):333–335, September 2002.

[14] M.-C. F. Chang, V. P. Roychowdhury, L. Zhang, H. Shin, and Y. Qian. RF/Wireless for inter- and intra-chip communications. *Proceedings IEEE*, 89(4):456–466, April 2001.

[15] R. E. Collin. *Antennas and Radiowave Propagation*. New York, NY: McGraw-Hill, 1985.

[16] R. E. Collin. Limitations of the Thévenin and Norton equivalent circuits for a receiving antenna. *IEEE Antennas and Propagation Magazine*, 45(2):119–124, April 2003.

[17] R. Courant and D. Hilbert. *Methods of Mathematical Physics*, volume 2. New York, NY: John Wiley & Sons, 1966.

[18] T. A. Cruse. A direct formulation and numerical solution of the general transient elastodynamic problem. (ii). *Journal of Mathematical Analysis and Applications*, 22:341–355, 1968.

[19] T. A. Cruse and F. J. Rizzo. A direct formulation and numerical solution of the general transient elastodynamic problem. (i). *Journal of Mathematical Analysis and Applications*, 22:244–259, 1968.

[20] A. T. de Hoop. A modification of Cagniard's method for solving seismic pulse problems. *Applied Scientific Research*, B(8):349–356, 1960.

[21] A. T. de Hoop. A reciprocity relation between the transmitting and the receiving properties of an antenna. *Applied Scientific Research*, 19(1):90–96, June 1968.

[22] A. T. de Hoop. Time-domain reciprocity theorems for electromagnetic fields in dispersive media. *Radio Science*, 22(7):1171–1178, December 1987.

[23] A. T. de Hoop. Reciprocity, discretization, and the numerical solution of direct and inverse electromagnetic radiation and scattering problems. *Proceedings IEEE*, 79(10):1421–1430, October 1991.

[24] A. T. de Hoop. *Handbook of Radiation and Scattering of Waves*. London, UK: Academic Press, 1995.

[25] A. T. de Hoop. Reflection and transmission of a transient, elastic line-source excited sh-wave by a planar, elastic bounding surface in a solid. *International Journal of Solids and Structures*, 39:5379–5391, 2002.

[26] A. T. de Hoop. A time-domain uniqueness theorem for electromagnetic wavefield modeling in dispersive, anisotropic media. *Radio Science Bulletin*, 305:17–21, June 2003.

[27] A. T. de Hoop and G. de Jong. Power reciprocity in antenna theory. *Proceedings of the Institution of Electrical Engineers*, 121(10):594–605, October 1974.

[28] A. T. de Hoop, I. E. Lager, and V. Tomassetti. The pulsed-field multiport antenna system reciprocity relation and its applications – a time-domain approach. *IEEE Transactions on Antennas and Propagation*, 57(3):594–605, March 2009.

[29] A. T. de Hoop, M. Stoopman, W. A. Serdijn, and I. E. Lager. Equivalent Thévenin and Norton Kirchhoff circuits of a receiving antenna. *IEEE Antennas and Wireless Propagation Letters*, 12:1627–1629, 2013.

[30] A. Deutsch, R. S. Krabbenhoft, K. L. Melde, C. W. Surovic, G. A. Katopis, G. V. Kopcsay, Z. Zhou, Z. Chen, Y. H. Kwark, T.-M. Winkel, X. Gu, and T. E. Standaert. Application of the short-pulse propagation technique for broadband characterization of PCB and other interconnect technologies. *IEEE Transactions on Electromagnetic Compatibility*, 52(2):266–287, May 2010.

[31] J. Dominguez. *Boundary Elements in Dynamics*. Southampton, UK: Computational Mechanics Publications, 1993.

[32] X. Duan, R. Rimolo-Donadio, H.-D. Brüns, and C. Schuster. A combined method for fast analysis of signal propagation, ground noise, and radiated emission of multilayer printed circuit boards. *IEEE Transactions on Electromagnetic Compatibility*, 52(2):487–495, May 2010.

[33] A. P. Duffy, A. J. M. Martin, A. Orlandi, G. Antonini, T. M. Benson, and M. S. Woolfson. Feature selective validation (FSV) for validation of computational electromagnetics (CEM). Part I – the FSV method. *IEEE Transactions on Electromagnetic Compatibility*, 48(3):449–459, August 2006.

[34] D. G. Duffy. *Transform Methods for Solving Partial Differential Equations*. Boca Raton, FL: Chapman & Hall/CRC, 2nd edition edition, 2004.

[35] M. A. Elmansouri and D. S. Filipovic. Pulse distortion and mitigation thereof in spiral antenna-based UWB communication systems. *IEEE Transactions on Antennas and Propagation*, 59(10):3863–3871, October 2011.

[36] J. Fan, W. Cui, J. L. Drewniak, T. P. Van Doren, and J. L. Knighten. Estimating the noise mitigation effect of local decoupling in printed circuit boards. *IEEE Transactions on Advanced Packaging*, 25(2):154–165, May 2002.

[37] L. B. Felsen. Propagation and diffraction of transient fields in non-dispersive and dispersive media. In Felsen L. B., editor, *Transient Electromagnetic Fields*, chapter 1, pages 1–72. Berlin, Germany: Springer - Verlag, 1976.

[38] M. B. Friedman and R. Shaw. Diffraction of pulse by cylindrical obstacles of arbitrary cross section. *Journal of Applied Mechanics*, 29:40–46, 1962.

[39] M. Friedrich, M. Leone, and C. Bednarz. Exact analytical solution for the via-plate capacitance in multiple-layer structures. *IEEE Transactions on Electromagnetic Compatibility*, 54(5):1097–1104, 2012.

[40] F. Gardiol. Open question to time-domain experts. *IEEE Antennas and Propagation Society Newsletter*, 31(4):48, October 1988.

[41] R. Garg, P. Bhartia, I. Bahl, and A. Ittipiboon. *Microstrip Antenna Design Handbook*. Norwood, MA: Artech House, 2001.

[42] Q. Gu, Y. E. Yang, and M. A. Tassoudji. Modeling and analysis of vias in multilayered integrated circuits. *IEEE Transactions on Microwave Theory and Techniques*, 41(2):206–214, February 1993.

[43] K. C. Gupta. Multiport network approach for modelling and analysis of microstrip patch antennas and arrays. In J. R. James and P. S. Hall, editors, *Handbook of Microstrip Antennas*, volume 28 of *IEE Electromagnetic Waves*. Peter Peregrinus Ltd., 1989.

[44] A. Hardock, H.-D. Brüns, and C. Schuster. Chebyshev filter design using vias as quasi-transmission lines in printed circuit boards. *IEEE Transactions on Microwave Theory and Techniques*, 63(3):976–985, March 2015.

[45] R. F. Harrington. *Field Computation by Method of Moments*. Piscataway, NJ: IEEE Press, 2003.

[46] R. F. Harrington, K. Pontoppidan, P. Abrahamsen, and N. C. Albertsen. Computation of Laplacian potentials by an equivalent source method. *Proceedings IEEE*, 116(10):1715–1720, October 1969.

[47] J. D. Hoffman. *Numerical Methods for Engineers and Scientists*. New York, NY: Marcel Dekker, 2001.

[48] A. Ishimaru. *Electromagnetic Wave Propagation, Radiation and Scattering*. Englewood Cliffs, NJ: Prentice-Hall, Inc., 1991.

[49] D. R. Jackson, W. F. Richards, and A. Ali-Khan. Input impedance and mutual coupling of rectangular microstrip antennas. 37(3):269–274, March 1989.

[50] M. A. Jaswon. Integral equation methods in potential theory. (i). *Proceedings of the Royal Society of London*, 275:23–32, January 1963.

[51] F. F. Judd and P. M Chirlian. The application of the compensation theorem in the proof of Thevenin's and Norton's theorems. *IEEE Transactions on Education*, 2(13):87–88, 1970.

[52] M. Y. Koledintseva, J. L. Drewniak, D. J. Pommerenke, Orlandi A. Antonini, G., and K. N. Rozanov. Wide-band Lorentzian media in the FDTD algorithm. *IEEE Transactions on Electromagnetic Compatibility*, 47(2):392–399, May 2005.

[53] M. Koshiba and M. Suzuki. Application of the boundary-element method to waveguide discontinuities. *IEEE Transactions on Microwave Theory and Techniques*, 34(2):301–307, February 1986.

[54] I. E. Lager and A. T. de Hoop. Inter-chip and intra-chip pulsed signal transfer between transmitting and receiving loops in wireless interconnect configurations. In *Proc. 40th European Microwave Conference*, pages 577–580, Paris, France, September 2010.

[55] I. E. Lager, A. T. de Hoop, and T. Kikkawa. Pulsed-field wireless interconnects in digital integrated circuits – a time-domain signal transfer and electromagnetic emission analysis. In *Proc. 6th European Conference on Antennas and Propagation*, pages 1855–1859, Prague, The Czech Republic, March 2012.

[56] J.L. Lagos and F Fiori. Worst-case induced disturbances in digital and analog interchip interconnects by an external electromagnetic plane wave – Part I: modeling and algorithm. *IEEE Transactions on Electromagnetic Compatibility*, 53(1):178–184, February 2011.

[57] G.-T. Lei. Examination, clarification, and optimization of the Greens function/z-matrix models and calculations for rectangular planar microwave circuits. *IEEE Transactions on Microwave Theory and Techniques*, 59(4):803–815, April 2011.

[58] G.-T. Lei, R. W. Techentin, and B. K. Gilbert. High-frequency characterization of power/ground-plane structures. *IEEE Transactions on Microwave Theory and Techniques*, 47(5):562–569, May 1999.

[59] M. Leone. The radiation of a rectangular power-bus structure at multiple cavity-mode resonances. *IEEE Transactions on Electromagnetic Compatibility*, 45(3):486–492, August 2003.

[60] M. Leone. Radiated susceptibility on the printed-circuit-board level: simulation and measurement. *IEEE Transactions on Electromagnetic Compatibility*, 47(3):471–478, August 2005.

[61] Y. T. Lo, D. Solomon, and W. F. Richards. Theory and experiment on microstrip antenna. *IEEE Transactions on Antennas and Propagation*, 27(2):137–145, March 1979.

[62] S. Maeda, T. Kashiwa, and I. Fukai. Full wave analysis of propagation characteristics of a through hole using the finite-difference time-domain method. *IEEE Transactions on Microwave Theory and Techniques*, 39(12):2154–2159, December 1991.

[63] M. Malkomes. Mutual coupling between microstrip patch antennas. *Electronics Letters*, 18(12):520–522, June 1982.

[64] W. J. Mansur and C. A. Brebbia. Formulation of the boundary element method for transient problems governed by the scalar wave equation. *Applied Mathematical Modelling*, 6:307–311, August 1982.

[65] W. J. Mansur and C. A. Brebbia. Numerical implementation of the boundary element method for two dimensional transient scalar wave propagation problems. *Applied Mathematical Modelling*, 6:299–306, August 1982.

[66] R. Mittra. A vector form of compensation theorem and its application to boundary-value problems. *Applied Scientific Research*, 11(1-2):26–42, 1964.

[67] R. Mittra and C. A. Klein. Stability and convergence of moment method solutions. In R. Mittra, editor, *Numerical and Asymptotic Techniques in Electromagnetics*. Berlin: Springer-Verlag, 1975.

[68] P. M. Morse and H. Feshbach. *Methods of Theoretical Physics, Part I*. McGraw-Hill, New York, 1953.

[69] P. M. Morse and H. Feshbach. *Methods of Theoretical Physics, Part II*. McGraw-Hill, New York, 1953.

[70] J. R. Mosig. Integral equation technique. In Itoh T., editor, *Numerical Techniques for Microwave and Millimeter-Wave Passive Structures*, chapter 3, pages 133–213. New York, NY: John Wiley & Sons, 1989.

[71] J. R. Mosig, R. C. Hall, and F. E. Gardiol. Numerical analysis of microstrip patch antennas. In J. R. James and P. S. Hall, editors, *Handbook of Microstrip Antennas*, volume 28 of *IEE Electromagnetic Waves*. Peter Peregrinus Ltd., 1989.

[72] T. Myioshi and S. Miyauchi. The design of planar circulators for wide-band operation. *IEEE Transactions on Microwave Theory and Techniques*, 28(3):210–214, July 1977.

[73] T. Myioshi, S. Yamaguchi, and S. Goto. Ferrite planar circuits in microwave integrated circuits. *IEEE Transactions on Microwave Theory and Techniques*, 25(7):593–600, July 1977.

[74] E. H. Newman and D. Forrai. Scattering from a microstrip patch. *IEEE Transactions on Antennas and Propagation*, 35(3):245–251, March 1987.

[75] E. H. Newman, J. H. Richmond, and B. W. Kwan. Mutual impedance computation between microstrip antennas. *IEEE Transactions on Microwave Theory and Techniques*, 31(11):941–945, November 1983.

[76] N. Nguyen-Trong, T. Kaufmann, L. Hall, and C. Fumeaux. Analysis and design of a reconfigurable antenna based on half-mode substrate-integrated cavity. *IEEE Transactions on Antennas and Propagation*, 63(8):3345–3353, August 2015.

[77] P. V. Nikitin and K. V. S. Rao. Theory and measurement of backscattering from rfid tags. *IEEE Antennas and Propagation Magazine*, 48(6):212–218, 2006.

[78] J. Nitsch, S. V. Tkatchenko, and S. Potthast. Transient excitation of rectangular resonators through electrically small circular holes. *IEEE Transactions on Electromagnetic Compatibility*, 54(6):1252–1259, December 2012.

[79] Y. Niwa, S. Hirose, and M. Kitahara. Application of the boundary integral equation method to transient analysis of inclusions in a half space. *Wave Motion*, 8:77–91, January 1986.

[80] T. Okoshi. *Planar Circuits for Microwaves and Lightwaves*. Springer Series in Electrophysics. Berlin: Springer-Verlag, 1985.

[81] T. Okoshi and S. Kitazawa. Computer analysis of short-boundary planar circuits. *IEEE Transactions on Microwave Theory and Techniques*, 23(3):299–306, March 1975.

[82] T. Okoshi and T. Miyoshi. The planar circuit - an approach to microwave integrated circuitry. *IEEE Transactions on Microwave Theory and Techniques*, 20(4):887–892, April 1972.

[83] T. Okoshi, Y. Uehara, and T. Takeuchi. The segmentation method - an approach to the analysis of microwave planar circuits. *IEEE Transactions on Microwave Theory and Techniques*, 24(10):662–668, October 1976.

[84] A. Orlandi, A. P. Duffy, B. Archambeault, G. Antonini, D. E. Coleby, and S. Connor. Feature selective validation (FSV) for validation of computational electromagnetics (CEM). Part II – assessment of FSV performance. *IEEE Transactions on Electromagnetic Compatibility*, 48(3):460–467, August 2006.

[85] J. C. Parker. Via coupling within parallel rectangular planes. *IEEE Transactions on Electromagnetic Compatibility*, 39(1):17–23, February 1997.

[86] C. A. Paul. *Introduction to Electromagnetic Compatibility*. New York, NY: John Wiley & Sons, Inc., 2nd edition edition, 2006.

[87] E. Penard and J.-P. Daniel. Mutual coupling between microstrip antennas. *Electronics Letters*, 18(14):605–607, July 1982.

[88] R. Porath. Theory of miniaturized shorting-post microstrip antennas. *IEEE Transactions on Antennas and Propagation*, 48(1):41–47, January 2000.

[89] D. M. Pozar. Input impedance and mutual coupling of rectangular microstrip antennas. 30(6):1191–1196, November 1982.

[90] D. M. Pozar. Radiation and scattering from a microstrip patch on a uniaxial substrate. *IEEE Transactions on Antennas and Propagation*, 35(6):613–621, June 1987.

[91] D. M. Pozar. Closed-form approximations for link loss in a UWB radio system using small antennas. *IEEE Transactions on Antennas and Propagation*, 51(9):2346–2352, September 2003.

[92] D. M. Pozar. Waveform optimizations for ultrawideband radio systems. *IEEE Transactions on Antennas and Propagation*, 51(9):2335–2345, September 2003.

[93] D. M. Pozar. Scattered and absorbed powers in receiving antennas. *IEEE Antennas and Propagation Magazine*, 46(1):144–145, February 2004.

[94] A. J. Pray, N. V. Nair, and B. Shanker. Stability properties of the time domain electric field integral equation using a separable approximation for the convolution with the retarded potential. *IEEE Transactions on Antennas and Propagation*, 60(8):3772–3781, August 2012.

[95] J. B. Preibisch, A. Hardock, and C. Schuster. Physics-based via and waveguide models for efficient SIW simulations in multilayer substrates. *IEEE Transactions on Microwave Theory and Techniques*, 63(6):1809–1816, June 2015.

[96] D. Quak. Analysis of transient radiation of a (traveling) current pulse on a straight wire segment. In *Proc. 2001 IEEE EMC International Symposium*, pages 849–854, Montreal, Que., Canada, July 2001.

[97] P. J. Restle, A. E. Ruehli, S. G. Walker, and G Papadopoulos. Full-wave peec time-domain method for the modeling of on-chip interconnects. *IEEE Transactions on Computer-Aided Design of Integrated Circuits and Systems*, 20(7):877–886, July 2001.

[98] R. Rimolo-Donadio, X. Gu, Y. H. Kwark, M. B. Ritter, B. Archambeault, F. De Paulis, Y. Zhang, J. Fan, H.-D. Brüns, and C. Schuster. Physics-based via and trace models for efficient link simulation on multilayer structures up to 40 GHz. *IEEE Transactions on Microwave Theory and Techniques*, 57(8):2072–2083, August 2009.

[99] C. R. Rowell and R. D. Murch. A capacitively loaded PIFA for compact mobile telephone handsets. *IEEE Transactions on Antennas and Propagation*, 45(5):837–842, May 1997.

[100] V. H. Rumsey. Reaction concept in electromagnetic theory. *Physical Review*, 94(6):1483–1491, June 1954.

[101] K. Sakai and M. Koshiba. Analysis of electromagnetic field distribution in tunnels by the boundary-element method. In *IEE Proceedings H, Microwaves, Antennas and Propagation*, volume 137, pages 202–208, August 1990.

[102] T. K. Sarkar and S. M. Rao. The application of the conjugate gradient method for the solution of electromagnetic scattering from arbitrarily oriented wire antennas. *IEEE Transactions on Antennas and Propagation*, 32(4):398–403, 1984.

[103] M. Schanz and H. Antes. Application of 'Operational Quadrature Methods' in time domain boundary element methods. *Meccanica*, 32:179–186, 1997.

[104] J. P. Schouten. A new theorem in operational calculus together with an application of it. *Physica*, 2:75–80, 1935.

[105] T. E. Shea. *Transmission Networks and Wave Filters*. New York, NY: van Nostrand, 1929.

[106] J. Shim, D. G. Kam, J. H. Kwon, and J. Kim. Circuital modeling and measurement of shielding effectiveness against oblique incident plane wave on apertures in multiple sides of rectangular enclosure. *IEEE Transactions on Electromagnetic Compatibility*, 52(3):566–577, August 2010.

[107] A. Shlivinski. Time-domain transfer coupled response of antennasreciprocity theorem approach. *IEEE Transactions on Antennas and Propagation*, 65(4):1714–1727, April 2017.

[108] A. Shlivinski, E. Heyman, and R. Kastner. Antenna characterization in the time domain. *IEEE Transactions on Antennas and Propagation*, 45(7):1140–1149, July 1997.

[109] J. C. Slater. *Microwave Electronics*. New York, NY: D. Van Nostrand Company, Inc., 1950.

[110] J. R. Solin. Formula for the field excited in a rectangular cavity with a small aperture. *IEEE Transactions on Electromagnetic Compatibility*, 53(1):82–90, February 2011.

[111] J. A. Stratton. *Electromagnetic Theory*. McGraw-Hill, New York, 1941.

[112] G. T. Symm. Integral equation methods in potential theory. (ii). *Proceedings of the Royal Society of London*, 275:33–46, January 1963.

[113] C. T. Tai. A critical study of the circuit relations of two distant antennas. *IEEE Antennas and Propagation Magazine*, 44(6):32–37, December 2002.

[114] M. Tanaka and T. Matsumoto. Transient elastodynamic boundary element formulations based on the time-stepping scheme. *International Journal of Pressure Vessels and Piping*, 42:317–332, February 1990.

[115] M. Tanaka, H. Tsuboi, F. Kobayashi, and T Misaki. Transient eddy current analysis by the boundary element method using fourier transforms. *IEEE Transactions on Magnetics*, 29(2):1722–1725, March 1993.

[116] F. M. Tesche, M. V. Ianoz, and T. Karlsson. *EMC Analysis Methods and Computational Models*. New York, NY: John Wiley & Sons, Inc., 1997.

[117] A. G. Tijhuis. *Electromagnetic Inverse Profiling: Theory and Numerical Implementation*. Utrecht, the Netherlands: VNU Science Press, 1987.

[118] A. G. Tijhuis. Iterative techniques for the solution of integral equations in transient electromagnetic scattering. *Progress In Electromagnetics Research*, 5:455–538, 1991.

[119] J. Trinkle and A. Cantoni. Impedance expressions for unloaded and loaded power ground planes. *IEEE Transactions on Electromagnetic Compatibility*, 50(2):390–398, May 2008.

[120] B. Van der Pol. A theorem on electrical networks with an application to filters. *Physica*, 1:521–530, 1934.

[121] V. Vahrenholt and M. Leone. Efficient Foster-type macromodels for rectangular planar interconnections. *IEEE Transactions on Electromagnetic Compatibility*, 2(10):1686–1695, October 2012.

[122] J. G. van Bladel. *Electromagnetic Fields*. John Wiley & Sons, Hoboken, NJ, 2nd ed. edition, 2007.

[123] P. M. van den Berg. Iterative computational techniques in scattering based upon the integrated square error criterion. *IEEE Transactions on Antennas and Propagation*, 32(10):1063–1071, 1984.

[124] M. Štumpf. An application of the Cagniard-DeHoop technique for solving initial-boundary value problems in bounded regions. *The Quarterly Journal of Mechanics and Applied Mathematics*, 66(2):185–197, 2013.

[125] M. Štumpf. Pulsed EM field radiation, mutual coupling and reciprocity of thin planar antennas. *IEEE Transactions on Antennas and Propagation*, 62(8):3943–3950, August 2014.

[126] M. Štumpf. Time-domain analysis of rectangular power-ground structures with relaxation. *IEEE Transactions on Electromagnetic Compatibility*, 56(5):1095–1102, October 2014.

[127] M. Štumpf. The time-domain contour integral method – an approach to the analysis of double-plane circuits. *IEEE Transactions on Electromagnetic Compatibility*, 56(2):367–374, April 2014.

[128] M. Štumpf. Time-domain mutual coupling between power-ground structures. In *Proc. 2014 IEEE EMC & SIPI International Symposium*, pages 240–243, Raleigh, North Carolina, August 2014.

[129] M. Štumpf. Analysis of dispersive power-ground structures using the time-domain contour integral method. *IEEE Transactions on Electromagnetic Compatibility*, 57(2):224–231, April 2015.

[130] M. Štumpf. The pulsed EM plane-wave response of a thin planar antenna. *Journal of Electromagnetic Waves and Applications*, 30(9):1133–1146, 2016.

[131] M. Štumpf. The equivalent Thévenin-network representation of a pulse-excited power-ground structure. *IEEE Transactions on Electromagnetic Compatibility*, 59(1):249–255, February 2017.

[132] M. Štumpf. Extending the validity of the time-domain contour integral method using the admittance-wall boundary condition. In *Proc. 2017 IEEE EMC & SIPI International Symposium*, pages 751–755, Washington, D.C, August 2017.

[133] M. Štumpf. *Electromagnetic Reciprocity in Antenna Theory*. Hoboken, NJ: Wiley–IEEE Press, 2018.

[134] M. Štumpf. The time-domain compensation theorem and its application to pulsed EM scattering of multiport receiving antennas. *IEEE Transactions on Antennas and Propagation*, 66(1):226–232, January 2018.

[135] M. Štumpf and I. E. Lager. The time-domain optical theorem in antenna theory. *IEEE Antennas and Wireless Propagation Letters*, 14:895–897, 2015.

[136] M. Štumpf and M. Leone. Efficient 2-D integral equation approach for the analysis of power bus structures with arbitrary shape. *IEEE Transactions on Electromagnetic Compatibility*, 51(1):38–45, February 2009.

[137] M. Štumpf and Z. Raida. Pulsed electromagnetic waves between parallel plates: the modal-expansion and generalized-ray approaches. *IEEE Antennas and Propagation Magazine*, 56(6):90–101, December 2014.

[138] J.-S. Wang and R. Mittra. Finite element analysis of MMIC structures and electronic packages using absorbing boundary conditions. *IEEE Transactions on Microwave Theory and Techniques*, 42(3):441–449, March 1994.

[139] X.-C. Wei, E.-P. Li, E.-X. Liu, and X. Cui. Efficient modeling of rerouted return currents in multilayered power-ground planes by using integral equation. *IEEE Transactions on Electromagnetic Compatibility*, 50(3):740–743, August 2008.

[140] D. Wilton, S. Rao, A. Glisson, D. Schaubert, O. Al-Bundak, and C. Butler. Potential integrals for uniform and linear source distributions on polygonal and polyhedral domains. *IEEE Transactions on Antennas and Propagation*, 32(3):276–281, March 1984.

[141] S. A. Winkler, W. Hong, M. Bozzi, and K. Wu. Polarization rotating frequency selective surface based on substrate integrated waveguide technology. *IEEE Transactions on Antennas and Propagation*, 58(4):1202–1213, April 2010.

[142] K.-B. Wu, R.-B. Wu, and D. De Zutter. Modeling and optimal design of shorting vias to suppress radiated emission in high-speed alternating PCB planes. *IEEE Transactions on Components, Packaging and Manufacturing Technology*, 1(4):566–573, April 2011.

[143] T.-L. Wu and S.-T. Chen. A photonic crystal power/ground layer for eliminating simultaneously switching noise in high-speed circuit. *IEEE Transactions on Microwave Theory and Techniques*, 54(8):3398–3406, August 2006.

[144] F. Xu, Y. Zhang, W. Hong, K. Wu, and T. J. Cui. Finite-difference frequency-domain algorithm for modeling guided-wave properties of substrate integrated waveguide. *IEEE Transactions on Microwave Theory and Techniques*, 51(11):2221–2227, November 2003.

[145] S. Yan, P. J. Soh, and G. A. E. Vandenbosch. Dual-band textile MIMO antenna based on substrate-integrated waveguide (SIW) technology. *IEEE Transactions on Antennas and Propagation*, 63(11):4640–4647, November 2015.

[146] E. Zauderer. *Partial Differential Equations of Applied Mathematics, 2nd Ed.* New York, NY: John Wiley & Sons, Inc., 1989.

[147] Y. Zhang, J. Fan, G. Selli, M. Cocchini, and F. de Paulis. Analytical evaluation of via-plate capacitance for multilayer printed circuit boards and packages. *IEEE Transactions on Microwave Theory and Techniques*, 56(9):2118–2128, 2008.

Index